# North American Mixing Forum 2016

Core Programming Area at the 2016 AIChE Annual Meeting

San Francisco, California, USA
13 - 18 November 2016

ISBN: 978-1-5108-3442-2

**Printed from e-media with permission by:**

Curran Associates, Inc.
57 Morehouse Lane
Red Hook, NY 12571

CURRAN ASSOCIATES INC. proceedings .com

**Some format issues inherent in the e-media version may also appear in this print version.**



Printed by Curran Associates, Inc. (2017)

For permission requests, please contact AIChE
at the address below.

AIChE
120 Wall Street, FL 23
New York, NY 10005-4020

Phone: (800) 242-4363
Fax: (203) 775-5177

www.aiche.org

**Additional copies of this publication are available from:**

Curran Associates, Inc.
57 Morehouse Lane
Red Hook, NY 12571 USA
Phone: 845-758-0400
Fax: 845-758-2633
Email: curran@proceedings.com
Web: www.proceedings.com

# TABLE OF CONTENTS

# 470016 Power Numbers and Power Correlations for a Retreat-Blade Impeller Under Different Baffling Conditions

**Abstract:**

Glass-lined tanks and reactors are commonly used in the in pharmaceutical API synthesis because of their corrosion resistance, cleanability, and reduced contamination level. A retreat-blade impeller with a low impeller clearance off the tank bottom is commonly used in glass-lined reactors provided with a torispherical vessel bottom. The power, P, dissipated by the impeller is critical to many mixing processes to achieve the desired mixing effect, especially since the power per unit volume, P/V, controls directly mass transfer processes and other mixing phenomena. Despite their common use in industry little information is available on power dissipation and the corresponding power number Po for this type of system is available in the literature.

In this study the power, P, dissipated by a retreat-blade impeller in different fluids of different viscosities at different agitation speeds and the non-dimensional Power Number, Po, were obtained experimentally in a scaled-down version of a typical glass-lined reactor for pharmaceutical API synthesis. Additionally, P and Po were obtained for an unbaffled system and a fully baffled system (four rectangular baffles) and for a large range of the Reynolds Number ($1<Re<1,000,000$). Po was found to vary significantly with baffling type and Re.

Equations correlating Po with Re and baffling type were obtained. These equations can be used by industrial users to help quantify and optimize pharmaceutical mixing processes typically involving the synthesis of APIs in this type of reactors.

# 472688 Turbulence Transition in Orbiting Culture Dishes

**Abstract:**

Orbiting culture dishes are widely used in the cell culture industry because they can deliver pulsatile biaxial flow while allowing for tens or hundreds of test cases to be run simultaneously. The orbiting dish system provides a means of mixing the fluid containing nutrients while simultaneously applying shear stress to cells in culture. Wall shear stresses (WSS) are widely accepted as the primary influence affecting characteristics of anchored cells subjected to fluid flow. Detailed, accurate information about the fluid forces acting on cells must be known in order to understand the cause and effect relationship between shear stresses and cellular responses, such as proliferation rates, morphology, and gene expression.

Computational fluid dynamics models describe how the traveling waveform in the dish was influenced by dimensionless Froude and Stokes numbers as well as a slope parameter defined as the ratio of the steady-state acceleration-induced free surface slope to the static fluid aspect ratio. The computational model provided complete spatial and temporal resolution of WSS over the bottom surface of a dish throughout the orbital cycle. Modeling the fluid dynamics in these orbiting dishes has always assumed laminar flow. Accurate experimental measurements have been very recently obtained using particle image velocimetry, which indicate that biologically relevant experimental conditions are in fact turbulent in many cases. Flow regime was determined by examining turbulent kinetic energy and velocity fluctuations. Randomly fluctuating velocities indicate turbulence while relatively stable velocities indicate laminar flow. Transition from laminar to turbulent flow was then defined and correlated with the above dimensionless parameters. The Froude number best described the transition to turbulence

# 467195 Numerical Simulation of Viscoelastic Liquid Flow in Stirred Vessels

**Abstract:**

Most of the fluids encountered in chemical, food or cosmetic industries exhibit non-Newtonian flow behaviour. Especially when dealing with polymeric and biological fluids, viscoelastic effects can occur, altering flow patterns and dynamics in comparison to Newtonian fluids.

In general, viscoelastic fluids can be viewed as a combination of a viscous and an elastic material. In contrast to Newtonian fluids, time-dependent stress relaxation is observed. When subjected to shearing, viscoelastic fluids develop normal stress differences, leading to phenomena like the Weissenberg effect. Furthermore, stretching of these fluids causes elastic forces. Additionally, the viscosity dependence on shear rate is commonly non-linear.

In stirred tanks, these effects can lead to cavern formation and inversion of flow fields around impellers i.e., influencing mixing times, power input, shear stress, etc.. For equipment design, the ability to quantify and predict these effects numerically is therefore of high interest in order to be able to choose optimal stirrer and tank geometries and process conditions.

Due to their complex flow behaviour, numerical simulation of viscoelastic fluids is a challenging task, especially when rotating geometries are considered. In most CFD software, it has not been possible to do such simulations yet. Hence, a tool for calculating viscoelastic flows with rotating geometries based on the Finite-Volume-Method in OpenFOAM was developed. By combining the split-stress tensor approach and viscoelastic differential constitutive equations with the sliding-mesh technique, it is now possible to perform simulations for viscoelastic fluids stirred by complex, rotating geometries.

With this tool, simulations were run for multiple stirrer geometries at various stirring rates and for different constitutive equations, using aqueous solutions of Xanthan and Carboxymethyl cellulose as model fluids. Since an exact description of the rheological properties of viscoelastic fluids is essential for quantitatively accurate flow simulations, rheological measurements of these solutions were conducted and material parameters for different constitutive equations were determined. For validation, the simulations were compared to flow field data acquired through particle image velocimetry measurements.

# 453071 Numerical Investigation of Mixing Performance of a High Viscous Fluid in the Novel Static Mixers with Different Group of Helical Leaves

**Abstract:**

The flow performance and mixing characteristic of a high-viscosity fluid in static mixers with radial multitwisted and axial twisted tapes were numerically investigated based on Lagrangian tracking method. The chaotic mixing performance was evaluated by the tracer particle distribution maps, extensional efficiency and stretching rate. The fluid in the three twisted leaves (TKSM) and four twisted leaves (FKSM) reached chaotic flow more quickly compared with KSM, and the mixing-element groups arranged in the beginning of the mixers played a main role in improving the dispersive mixing whereas the successive ones mainly maintained the dispersive mixing and improved the distribution performance. In contrast, there were large unmixed zones in the static mixer with DKSM. The results for the G values indicate that the macro-mixing performances in the DKSM are better than those in the KSM. KSM, MSM, and SSM had nearly identical stretching rate which were much higher than RSM for $Re<10$. With the increase of Re, the mixing performance of MSM decreased first and then increased which was just a little larger than that in RSM with the weakest micro-mixing ability. For a given length of static mixer, a decrease in aspect ratio benefited an increasing stretching rate.

# 451945 Evaluation of Change-Can Mixer Designs Via CFD Modeling

**Abstract:**

Change-can mixers are commonly used for processing high viscosity fluids at industrial scale due to their versatility. Multiple low and high shear blades that are independently driven and controlled can be used in mixing tank which makes it ideal for multitasking, to streamline processing operations and lower production costs. In order to obtain the desired properties, key process parameters require careful control, such as shearing intensity and mixing time. Predicting the mixing performance, such as mixing time, shear rate and fluid velocity, remains of key importance in process control, and it is the only way to ensure a high quality and uniform product.

To have a better understanding of the mixing impact of change-can mixers for producing stable mixtures, this study investigates the liquid-liquid mixing efficiency of different change-can mixer designs using the same impeller and disperser blade type. Computational fluid dynamics (CFD) modeling can provide detailed information about hydrodynamics and mixing. However, it is computationally intensive, especially multiphase transient simulations. Therefore, in this work, to reduce time and computational expense, the rotation of the blades in the change-can was modeled using the steady state multiple reference frames (MRF) technique with k– model. Subsequently, the mixing time analysis was calculated by the transient species transport model using fully developed flow field with commercial code Ansys Fluent 16.2.

Water and glycerin mixing was used for a case study. Three-dimensional (3D) simulations of mixers were created and compared: 1) the blades rotate at fixed positions, 2) the blades rotate at a planetary motion and 3) a spiral ribbon addition on the fixed blade. Simulations show faster mixing time with glycerin on top compared to water on top, and significant time improvement with a spiral ribbon addition on the fixed blade. In conclusion, a less expansive spiral ribbon addition on the fixed blade is more efficient as compared to a more expensive investment involving blades rotating in planetary motion.

**461342 Mixing of Non-Newtonian Fluids for Non-Aqueous Oral Health Products: Experimental Investigations and CFD Simulations**

**Abstract:**

Introduction

Mechanically stirred vessels are widely used in a broad range of chemical process industries, and their purpose varies from mixing different materials, generating solids suspensions or enhancing heat and mass transfer. The geometric configuration of both impeller and tank together with the fluid physical properties, in particular viscosity, dramatically affect the fluid dynamics of the system. For Newtonian fluids general guidelines for the most appropriate geometric configuration in different applications are available in the literature. Conversely, even if non-Newtonian fluids are employed in many industrial processes, less information is available. These fluids are complicated to mix, and often sophisticated impeller designs are required. This is the case in the oral health industry, where highly viscous non-Newtonian fluids are used in the production of non-aqueous specialized oral care products, such us non-aqueous toothpastes.

Computational fluid dynamics (CFD) is a powerful tool that yields relevant process information and permits assessing the performance of stirred vessels. The software relies on basic balance equations of fluid mechanics, such as mass, linear momentum and energy balance equations, solved on discretized fluid domains defined by the user. CFD can be used, among others, to evaluate the mixing performance of stirred tanks, which is particularly useful when dealing with non-Newtonian fluids. The objective of this research is twofold: we first aim to characterise the rheology of the two-fluid mixture of interest (glycerol and a polymer made of polyethylene glycol and Carbomer) and to derive a constitutive equation to model it. The second objective is to use this rheological information to numerically simulate the fluid dynamic behaviour of the mixture in a stirred vessel and to validate the model predictions against accurate experimental results of power consumption.

Experimental setup and methodology

The fluids of interest in this research are glycerol and a polymer made of polyethylene glycol and Carbomer at a specific ratio. The rheology of the polymer has not been studied before. For the development of the CFD model for the mixing of these fluids we need to determine the constitutive equation that describes the viscosity of mixtures of this particular polymer with glycerol at different polymer mass fractions and shear rates at different temperatures.

We studied the polymer mass fraction in intervals of 10%, from 0 to 100%. Each mixture was studied at thirteen equally-spaced temperatures, between 25 °C and 85 °C, according to the operating conditions of the industrial mixing process. For each polymer concentration and temperature considered, the viscosity of the mixture was evaluated at fifteen equidistant points in a logarithmic scale of shear rate in the range of 1 $s^{-1}$ to 250 $s^{-1}$, where the experimental results were repeatable. Indicative results are shown in Figure 1, which correspond to low and high values of polymer mass fraction (40% and 100%) and to low and high values of temperature (40°C and 60°C). We used the rheological data to determine the coefficients of a power law model for each experimental case (Eq. 1).

$$\mu = K\dot{\gamma}^{n-1} \quad \text{(Eq. 1)}$$

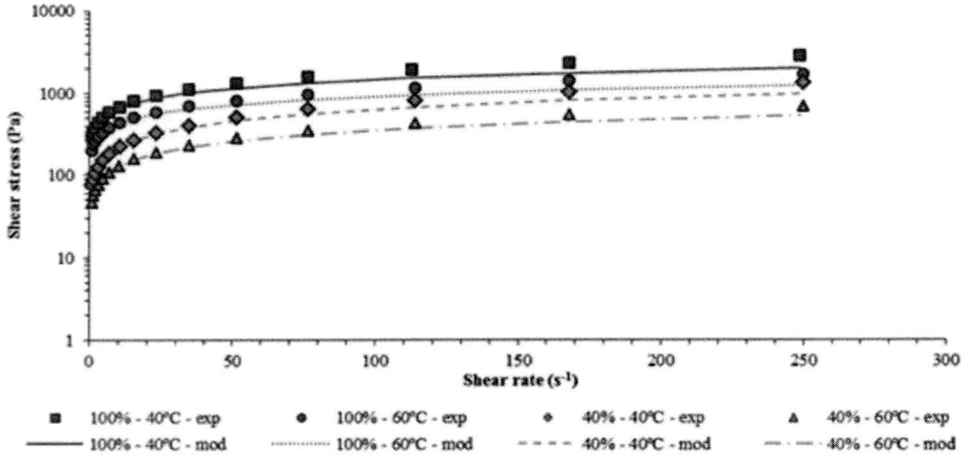

Figure 1. Shear stress profiles of samples with different polymer mass fractions and temperatures

This same approach was followed for all conditions studied, and a total of 143 values for each coefficient, *K* and *n*, were determined, which correspond to the 11 polymer mass fractions and 13 temperatures evaluated. The coefficients can be calculated at any value of polymer mass fraction from 0 to 1 and temperatures from 25 to 85 °C through linear interpolation.

Experimental measurements of power consumption

Power requirements for the mixing of the polymer/glycerol solutions were measured experimentally at certain concentrations and temperatures in order to validate the CFD model of the stirred tank with the viscosity function developed from the rheological measurements. The measurements were carried out in a stirred tank fitted with a standard Rushton impeller. The height of the fluid H was set equal to the tank diameter. The impeller has a diameter D=1/3H; it was located at the centre of the tank, and was driven by a variable-speed motor that could operate in the range of 50 – 2000 rpm. To measure power consumption we used a combination of an air bearing and a load cell (Figure 2).

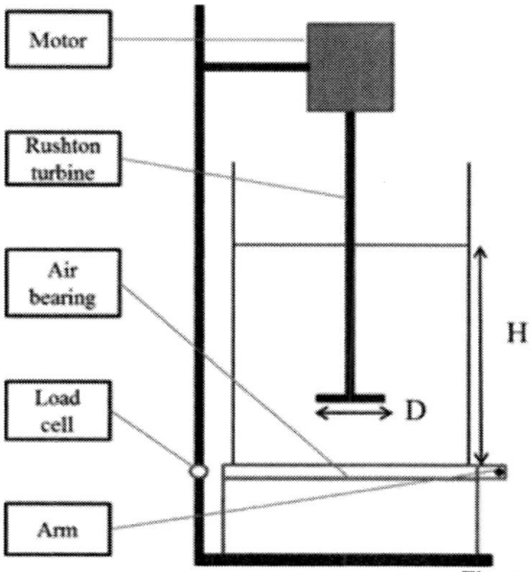


Figure 2. Schematics of the experimental setup

CFD modelling strategy

The two main approaches for modelling stirred tanks are the Reference Frame (RF) and the Sliding Mesh (SM). The main advantage of RF is the speed of convergence, but it can only be used for steady-state flows. In contrast, SM provides a time dependent solution, but the computational effort and time required to perform SM simulations are quite significant. RF is therefore the preferred approach when its predictions are the same as those given by the SM approach. To decide what approach to use, we tested both of them by modelling the mixing of glycerol at 25 °C at 100 rpm impeller speed in the set up described above. It was found that the RF approach is applicable. The computations were carried out using a 3.50 GHz Intel® Xeon ® CPU E5-1650 v2 with 16.0 GB RAM. The fluid dynamics were modelled using ANSYS Fluent 16.1. Results and Discussion

The torque and power consumption were measured for the mixing of five non-Newtonian glycerol/polymer mixtures at six impeller speeds and two different temperatures. In Figures 3 and 4, the power curves (plot of Power number against Reynolds number) are plotted for two different cases, 100% polymer mass fraction at 40 °C and 40% polymer mass fraction at 60 °C. In the same Figures, the CFD predictions are also included.

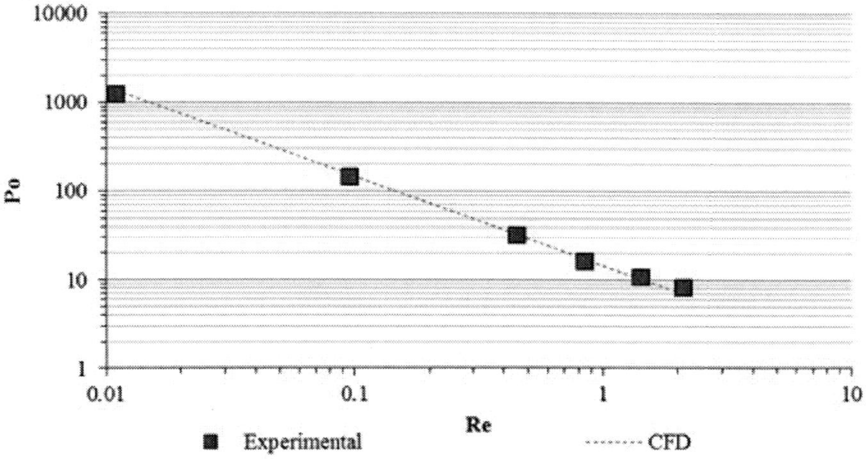

Figure 3. Power curve for 100% gel mass fraction and 40 °C

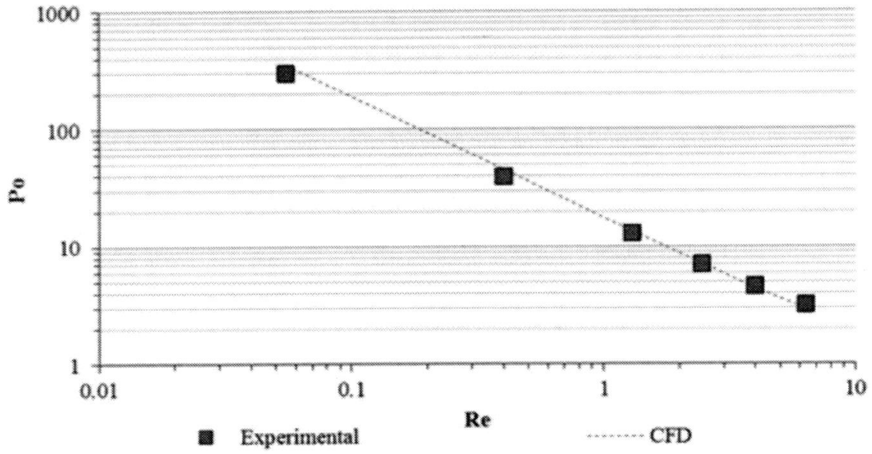


Figure 4. Power curve for 40% gel mass fraction and 60 °C

The agreement between the computational and experimental data is very good, which indicates that the CFD model can be used to predict the torque applied by a Rushton turbine using the experimental rheology model. Similar results are found for the remaining cases.

We currently study the mixing time of the two fluids both experimentally and computationally. For these studies the tank is filled with glycerol and the polymer is added at a particular location. Experimentally the concentration profiles in the tank are monitored with planar laser induced fluorescence (PLIF).

Acknowledgements

The authors would like to express their gratitude to GSK and the EPSRC for the funding to carry out this research.

# Numerical Simulation of the Turbulence Flow in Four Nozzles Opposed Impinging Stream Reactor

*Jianwei Zhang, Junjie Yan, Panlong Shang and Ying Feng*
*College of Energy and Power Engineering,*
*Shenyang University of Chemical Technology, SY, CHN*

## Abstract

In order to discuss the turbulence properties of the flow field in the multiple impinging stream reactor, the flow field of four nozzles opposed impinging stream reactor is simulated using Fluent software. Numerical simulations are performed in 3D chamber, by which the k-ε RNG model is used as turbulence model. The distributions of turbulence parameters (turbulence intensity and turbulence kinetic energy) are studied under different operational conditions with variable nozzles diameter and inlet flow rate. It shows that turbulence intensity and turbulence kinetic energy of the impingement region increase more significantly than that of region without impact, and the optimal conditions of microcosmic mixture are achieved. The distributions of turbulence kinetic energy are approximate to that of the turbulence intensity on the axial direction of the nozzles.

## Introduction

The concept of impinging stream has been presented [1] since the early sixties of last century and it drew lots of attentions for its internal flow characteristics of enhancement of mass and heat transfer in chemical engineering industry [2]. The existing literature focuses on mixture and flow field characteristics and material preparation of impinging stream mixer. The impinging stream mixer can promote the flow characteristics of micro mixing [3] and the impact of excellent performance for preparation of ultra fine powders in recent research [4]. The computational fluid dynamics (Computer Fluid Dynamics CFD) technology has been rapid development, combination of experiment and simulation, can be a deeper understanding of the flow in the reactor and reduce the calculation. Unger et al.[5] using PIV and CFD simulation method to study the fluid flow characteristics of different viscosities in impinging stream reactor (CIJR). Icardi et al. [6] compared with the simulation and experimental results of LES and IES model is also deeply studied.

In order to improve the mixing efficiency and to improve the mixing efficiency, it is necessary to study the turbulence characteristics in impinging stream reactor. In this article, the effects of different flow inlet rates, nozzle diameters on the flow field turbulence performance were studied by taking turbulence intensity and turbulence kinetic energy.

---

This work is supported by National Natural Science Foundation of China (Grant 21476141), the Natural Science Foundation of Liaoning Province (Grant 201202174)

## Numerical Simulation

### *1. Computational Domain and Boundary Conditions*

This section provides a short description of the computational details as they were applied, in order to obtain a convergent solution of this flow.

The four nozzles opposed impinging jets have been modeled as a three dimensional block where surface of outlet is defined as outflow boundaries. Four velocity inlets with the geometries of circular were placed on the four opposite surfaces. (Fig. 1). This experiment was conducted for the condition of immersion, moreover, d and Q presented the nozzle diameter and the inlet flow rate respectively. In this paper, the fluid of the domain is water of atmospheric pressure, the nozzle diameter d was arranged at 8mm, 10mm and 12mm, and the inlet flow rate Q was set to 0.4m3/h, 0.5 m3/h and 0.6 m3/h. The pressure reference point is chosen as the center of the exit section, the pressure and velocity are coupled with the SIMPLEC algorithm, and the pressure is discretized using the standard form, and the momentum is discretized. Model as shown in Fig.2.

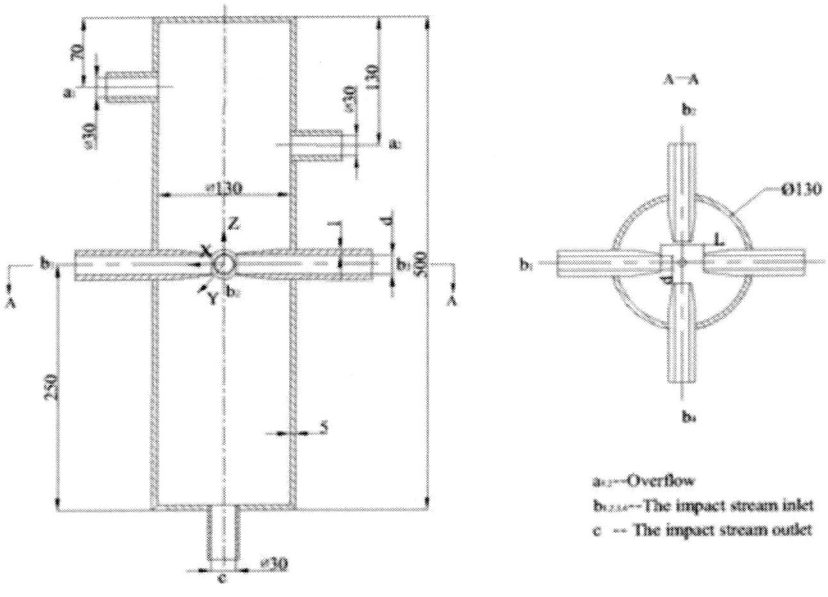


Fig. 1 Structure chart of impinging stream mixer

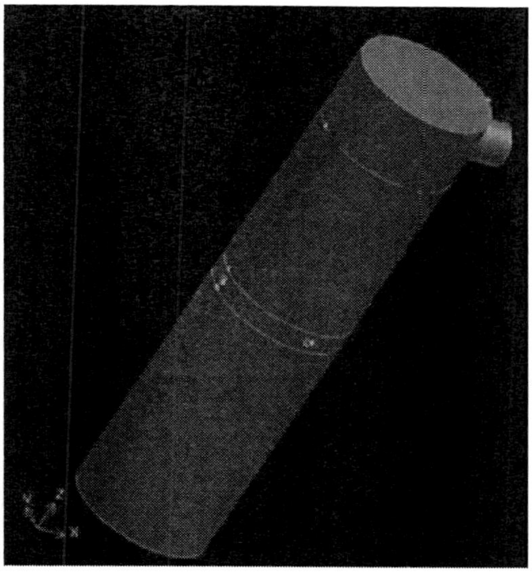

**Figure.2** model structure

### *2. CFD Modeling Strategy, Turbulence Model and Numerical Details*

The performance of the k-ε RNG model of turbulence has been investigated by Knowles et al. [7], Gordeev et al. [8], Sharif et al. [9], Fernandez et al. [10], Chem et al. [11], Liakos et al. [12,13], M. A. R. Sharif et al. [14] in impinging jet flows and all confirmed that the k-ε RNG model can be successfully used to model this type of complex flows. Thus the K-ε RNG turbulence model is used for modeling the turbulent behavior of the flow.

The first order upwind discretization scheme is applied for the momentum, turbulent kinetic energy and turbulent energy dissipation rate. SIMPLEST scheme has been used for the pressure–velocity coupling. Steady state simulations are carried out and the solution was considered to be converged when the sum of normalized residuals for continuity and momentum equations were on the order of $10^{-4}$.

### *3. Grid Generation*

An acceptable grid distribution is one in which grids are focused on important regions of domain. Regions surrounding the center line of the axial and radial jets are given the most importance. Therefore, the densest grid covers these areas. As on the axis of the jets, not only are the nozzle outlets important but also the impinging plane. The size of the cells being employed in the x direction is therefore uniform. Beyond the inlet region, a geometrically expanding grid is used with an expansion factor of 2 in both y and z directions.

## Results And Discussions

### *1. Effect of Different Nozzle Diameters on Velocity*

The distributions of velocity field on the y-z plane in the impinging stream mixer are depicted in **Figure 3**. The inlet flow rate Q is $0.5m^3/h$, the distance between nozzles distance is 3d, and the nozzle diameters are respectively 8mm, 10mm and 12mm.

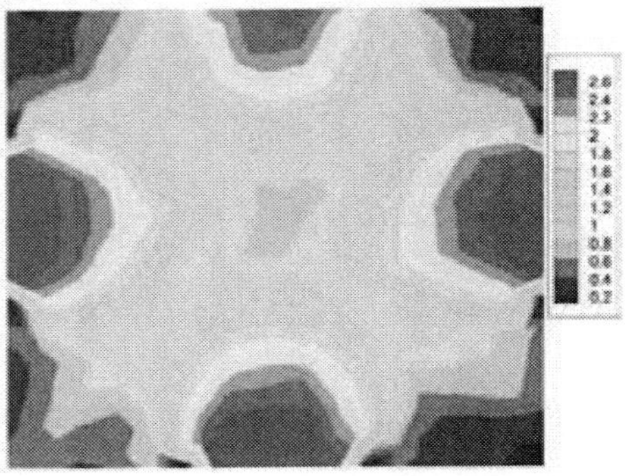

(a) d=8 mm

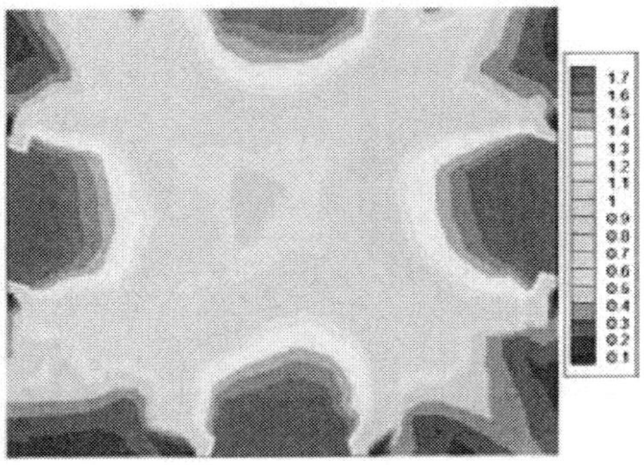

(b) d=10 mm

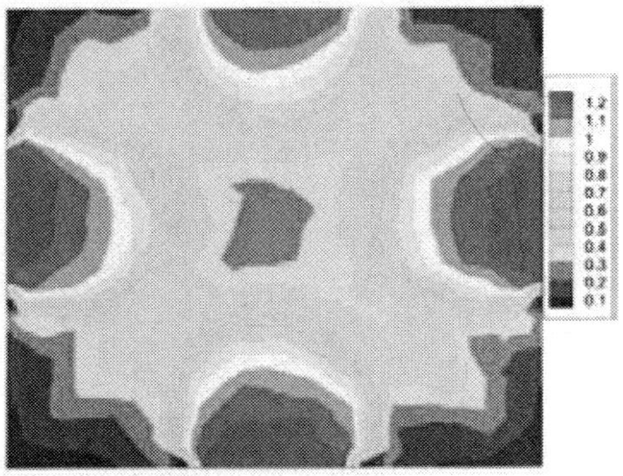

(c) d=12 mm

**Figure.3** The velocity field with different nozzle diameters on the y-z plane

The distributions of velocity are nearly centrosymmetric. The velocity decreases with the increase of the nozzle diameters. The values of velocity on the inlet are the maximum and the maximum values are respectively 2.6 m/s, 1.7 m/s and 1.2 m/s.The velocity is the smallest in the center area[15].

***2. Turbulence Characteristic Parameters***

The velocity of any point in the flow field is fluctuating under turbulent flow condition. The time average velocity represents kinematic velocity describing the velocity of a point in the turbulent field. In general, turbulence characteristic of flow field is defined by the parameters relative to velocity fluctuation, such as turbulence intensity and turbulence kinetic energy.

***2.1 Effect Of Different Nozzle Diameters On Turbulence Characteristic Parameters***

In order to specifically illustrate the influence of nozzle diameter on flow velocity field in the mixer, the turbulence characteristic parameters which is on the axis that includes the stagnation point of the impact region are analyzed.

The distributions of turbulence intensity and turbulence kinetic energy changing with the different nozzle diameters are shown in **Figure 4** (the nozzle diameter respectively is 8mm, 10mm and 12 mm).

High turbulence kinetic energy mainly concentrates on the impinging stagnation point. The peaks of two sides symmetrical on the stagnation point are the maximum that values are respectively

0.08 m$^2$/s$^2$, 0.035 m$^2$/s$^2$ and 0.02 m$^2$/s$^2$. These maximum values show a decreasing trend as the nozzle diameter increases. It is mainly due to the velocity fluctuation weakened by the increasing nozzle diameter.

As shown in **Figure 4** (b), the turbulence intensity distribution regularities of different nozzle diameters are similar and the turbulence intensity appears obvious changes at z-coordinate between –20 mm and 20 mm.In the direction away from the stagnation point, the changes of turbulent kinetic energy showed a decreased firstly and then increased.But the turbulence intensity changes in small range on the rest position. The turbulence intensity decreases as the nozzle diameter increases. The reinforcement of turbulence is reduced by the enlarging nozzle diameter, it is mainly due to weakened velocity fluctuation by increasing nozzle diameter.

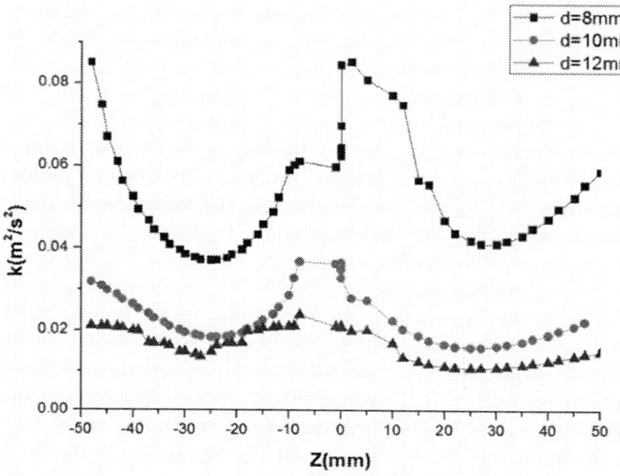


(a)Turbulence kinetic energy

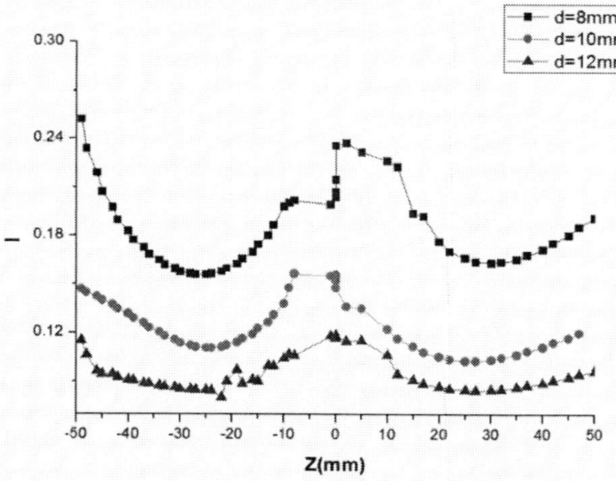


(b)Turbulence intensity

**Figure 4** The turbulence characteristic parameters with different inlet flow rates

*2.2 Effect of Different Inlet Flow Rates on Turbulence Characteristic Parameters*

In order to specifically illustrate the influence of inlet flow rates on flow velocity field in the mixer, the turbulence characteristic parameters which is on the axis that includes the stagnation point of the impact region are analyzed.

The distributions of turbulence intensity and turbulence kinetic energy changing with the different inlet flow rates are shown in **Figure 4** (the inlet flow rates Q respectively are 0.4m$^3$/h, 0.5m$^3$/h and 0.6m$^3$/h).

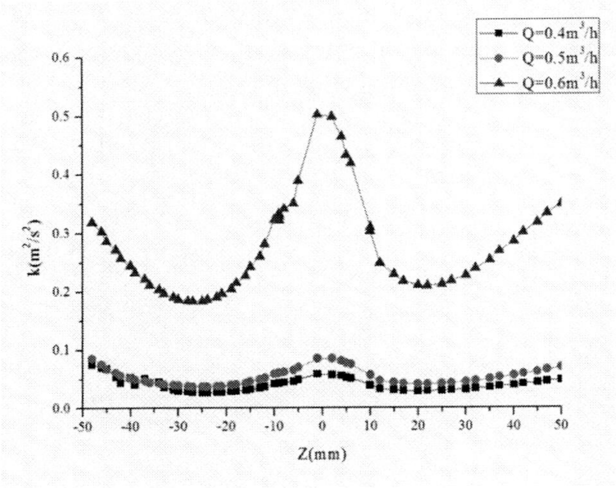


(a) Turbulence kinetic energy

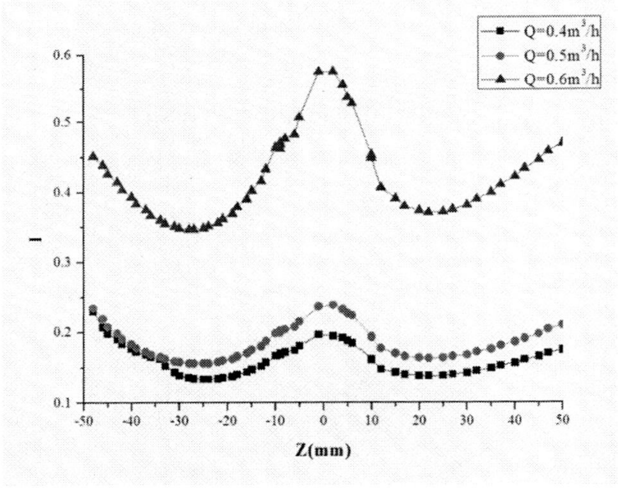


(b)Turbulence intensity

**Figure 4** The turbulence characteristic parameters with different inlet flow rates

Along the axis direction, the distribution of turbulent kinetic energy are mainly concentrated in the vicinity of the stagnation point, the values of turbulence kinetic energy is maximum in the impingement region, impact center was the most severe turbulence. When inlet flow rate Q is 0.6 m$^3$/h, the distribution of turbulent kinetic energy of in the center of the impact is in the shape of "positive triangle", and the shape is more regular. And in the flow of 0.4m$^3$/h and 0.5m$^3$/h, this change is not obvious. Mainly due to the velocity of inlet is not large. It is mainly caused by four fluids impact in the stagnation point. The interaction of shearing stress and pressure formed a large-scale vortex. It's beneficial to mixing.

As shown in Figure 4, the turbulence intensity distributions of different inlet flow rates are similar, the turbulence intensity appears obviously changes at coordinate-z between – 20 mm and 20 mm and the variation of which changes little on the rest position. The turbulence intensity is enhanced by increasing inlet flow rate. It is mainly caused by the enlarging velocity fluctuation as the inlet flow rate increases.

## Conclusions

The turbulence characteristics of the four nozzle opposed impinging stream mixer were discussed qualitatively by analyzing distributions of velocity, turbulence intensity and turbulence kinetic energy under different conditions (nozzle diameter and inlet flow rate change regularly). We can conclude as follow.

1. The velocity magnitude is greatly influenced by changing nozzle diameter of nozzles, in detail, the velocity reduces with the enlarging nozzle diameter.The corresponding characteristic parameters at stagnation point reach the minimum. The maximum values appear in the inlet.

2. The distributions of turbulence kinetic energy are approximate to that of the rms velocity. The turbulence kinetic energy reduces with the increasing nozzle diameter and increases with the enlarging inlet flow rate, which is affected most greatly by changing inlet flow rate. It can conclude that the increasing inlet flow rate enlarges the distribution area of high turbulence kinetic energy, which was beneficial to mixture.

3. The influence of changeable operating conditions on the turbulence intensity mainly reflected in the region of z=-20~20 mm, in which the corresponding values of turbulence intensity change obviously. In the text,the turbulence intensity changes the most greatly with the $Q=0.6$ $m^3/h$, and the turbulence intensity changes the least in case of the nozzle diameter d=12mm. The turbulence intensity reduces with increasing nozzle diameter between nozzles and increases with the enlarging inlet flow rate.

## Nomenclature

| | |
|---|---|
| d | inlet nozzle distance (mm) |
| L | inlet nozzle distance(mm) |
| u | velocity (m/s) |
| Q | inlet flow rate ($m^3/h$) |
| k | turbulence kinetic energy ($m^2/s^2$) |
| I | turbulence intensity |

## References


1. Elperin I T (1961). Heat and mass transfer in opposing currents[J]. *J Eng. Physics*, 62-68.

2. Tamir A, Kitron A(1987), Applications of Impinging Streams in Chemical Engineering Process-Review[J]. *Chem. Eng. Commun*, 241-330.
3. Yuan. Wu(2007), Impinging Streams: Fundamenals, Properties and Applications,Elsevier, Amsterdam, The Netherlands, 390
4. Yuan. Wu, Y. Zhou, J. Guo, J. Yuan(2010), Features of Impinging Streams Intensifying Processes and Their Applications, International Journal of Chemical Engineering, 1687-806X .
5. Unger D R, Muzzio F J, Brodkey R S(1998). Experimental and numerical characterization of viscous flow and mixing in an impinging jet contactor[J], The Canadian Journal of Chemical Engineering, 546-555.
6. Icardi M,Gavi E, Marchisio D L, et al. (2011),Validation of LES predictions for turbulent flow in a Confined Impinging Jets Reactor[J]. Applied Mathematical Modelling, 1591-1602
7. Knowles K, Saddington AJ (2002) ,Modelling and experiments on underexpanded turbulent jet mixing, 5th international symposium on engineering turbulence modelling and measurements, Mallorca, Spain, 16–18
8. Gordeev S, Heinzel V, Slobodchuk V (2006) ,Simulation of single and multiple impinging jet cooling and comparison with experimental data, Proceedings of the 8th biennial conference on engineering systems design and analysis, Turin, Italy, 04–07
9. Sharif MAR, Mothe KK (2009) Evaluation of turbulence models in the prediction of heat transfer due to slot jet impingement on plane and concave surfaces. Numerical Heat Transfer B Fundamation,273–294
10. Fernandez JA, Elicer-Cortes JC, Valencia A, et al. (2007) ,Comparison of low-cost two-equation turbulence models for prediction flow dynamics in twin-jets devices, 13th international heat transfer conference, Sydney, Australia, 13–18
11. Chen Q (1995) Comparison of different k-ε models for indoor air flow computations. Numer Heat Transfer B Fundamation ,353–369
12. Liakos HH, Koukou MK, Founti MA, Markatos NC(1998) Modeling of impinging turbulent jets and premixed flames, 4th European computational fluid dynamics conference, Athens, Greece, 07–11
13. Liakos HH, Keramida EP, Founti MA et al (2002) ,Heat and mass transfer study of impinging turbulent premixed flames. Heat and Mass Transfer,425–432
14. Sharif MAR, Mothe KK (2009),Parametric study of turbulent slot-jet impingement heat transfer from concave cylindrical surfaces. Int J Therm Sci ,428–442
15. Yongjiu Yu (2014). Experimental investigation and numerical simulation of flow characteristics in confined impinging jets reactor[D]. China，Beijing University of Chemical Technology.

# 462097 A Model of Droplet Dispersion in a Turbulent Couette Flow

**Abstract:**

A model of droplet dispersion in a turbulent flow in a Couette device in which the inner cylinder rotates whereas the outer cylinder is immobile, is developed. Flow regimes, characterized by relatively high Reynolds numbers (>13000), are modeled. Only dilute dispersions are considered. An evolution of droplet size distribution in a Couette device with time is described by an advection-diffusion equation containing a population balance term. Both droplet breakup and coalescence processes are taken into account. A one-dimensional engineering model of a turbulent Couette flow, based on the Prandtl mixing length concept, is employed for calculations of flow velocity, eddy diffusivity and turbulence energy dissipation rate distributions along a Couette device radius. The model equations are solved numerically. Size distributions of droplets, generated in a small-scale Couette device in presence of emulsion stabilizing surfactants, at different rotation speeds and operation durations are measured. The experimental data obtained are compared with the modeling results obtained at neglecting coalescence. The droplet dispersion process in absence of emulsion stabilizing chemicals is also illustrated by computational examples.

# 464847 The Relationship Between Contact Pressure and Coalescence Time for Water-in-Bitumen Emulsions

**Abstract:**

After the froth treatment section in the processing of oil sands from Alberta, the product – diluted bitumen, still contains about 2 to 3 wt% of water[1]. The salts in the residual water have strong implications for bitumen upgradation, including the possibility of corrosion of equipment and poisoning of catalysts, both of which lead to significant downtimes. In the past, coalescence has been evaluated as a strategy to combine the small water droplets with themselves or with bigger water droplets to produce larger drops that can be separated by inclined settling and centrifugation. Unfortunately, these studies have indicated that these water-in-bitumen emulsions are extremely difficult to break[1,2], and ascribe this difficulty to the presence of asphaltenes at the oil-water interface. A review of these studies revealed that there were at least two key deficiencies in these studies. First, very few of these studies were performed under industrially relevant conditions in terms of the weight percent of bitumen and the aromaticity of the diluent. Second, we believe that the contact times and contact forces were not sufficiently high to allow coalescence.

In this work, we have performed experiments with dueling, Hele-Shaw, water drops at the end of micropipettes within a microchannel, contacting them at controlled times and contact pressures. Our modified experimental setup allows us to study the coalescence of water drops in bitumen concentrations between 33 and 67 wt% (a solvent to bitumen ratio ranging between 0.5 to 2, which is close to the industrial operating range). Our investigations show that unlike prior studies, water drops in bitumen do coalesce, but only if the right combination of contact time and contact pressure are provided. The greater the contact pressure, the shorter the coalescence time. At a given contact pressure, the coalescence time increases as the weight percent of bitumen increases, but decreases as the pH of water is increased from 8.3 to 9.9. Empirical correlations are developed for the contact time as a function of the contact pressure for each combination of bitumen concentration and pH. These correlations are used to generate maps of drop radius vs. strain rate, outlining the parameter regimes in which coalescence should be expected in

sheared emulsions. The regime of drop radius vs. strain rate for coalescence expands with pH, as expected, but also expands as the bitumen concentration is increased, which is counterintuitive. These aspects will be explained in detail in the presentation.

**References**

1. F. Rao and Q. Liu, "Froth Treatment in Athabasca Oil Sands Bitumen Recovery Process: A Review", Energy Fuels **27**, 7199–7207 (2013).

Liquid Emulsion Films" , Langmuir **30**, 3024–3033 (2014).

# 470652 Mixing Assessments for Industrial Stirred TankÂ Reactors with Gas Entrainment from the Headspace

**Abstract:**

This mixing study was prompted by a difference in reaction rates in two semi-batch reactors that make the same product. The reactors are located at different production sites and have differing geometries. A gaseous reactant is introduced into the headspace of the stirred reactors containing a liquid phase co-reactant. The mixing is intended to provide a degree of contacting between the two reactant phases.

The pressure in the reactor headspace is sensitive to the consumption rate of the gaseous reactant, and thus provides a convenient metric for mass transfer behavior. The pressure data for the vessel with the slower reaction rate clearly shows long periods of poor contacting separated by short period of good mixing when liquid level occurs at impeller location. This is confirmed by motor amp readings. Laboratory experiments were performed to confirm the existing aeration behaviors and identify a suitable (i) impeller type, (ii) agitation speed, and (iii) number of impellers to improve surface aeration.

Significant capacity gains can be brought about in the sold-out reactor by improving the mixing performance. To give an estimate of potential, a reduction of < 30 minutes per batch equates to several million additional pounds of product per year (several percent gain in capacity).

Based on the plant data and experiments, a new agitation system was proposed. This design is currently moving toward implementation.

# 463341 Toward a Correlation for SOLID Liquid Cloud Height in Mechanical Agitated Vessels

**Abstract:**

Agitated vessels containing mixture of solids and liquids can be characterized by a minimum suspension speed and a vertical solids distribution. In many cases the cloud height which is the visual demarcation between slurry and clear interface is a useful parameter for characterizing vertical distribution. The emphasis in these studies has been on the effect of physical properties: concentration, density and viscosity and the effect of scale up. Geometry has been fixed throughout. The first thing noted was the insensitivity to physical properties. It is shown that the minimum suspension speed is not a correlating or normalizing variable. It was also shown that the relation between cloud height and speed was independent of whether the speed was above or below the minimum suspension speed. The appropriate normalizing value for cloud height was the tank diameter not the liquid level. Detailed studies were made of the effect of viscosity and concentration to confirm the weak functionalities. At lower concentrations there is no effect of concentration on cloud height.

The mechanism seems to be one of turbulent dispersion in the upper part of the vessel.

# 461833 Predicting Non-Mixing Regions of Granular Flows in a Spherical Tumbler

**Abstract:**

Mixing of granular materials is a key element in ensuring product uniformity in various industrial processes. However, the design of solids mixers has not changed significantly over the years due to under-developed theories and experimental approaches for studying three-dimensional (3D) granular flows. We consider the simple flow geometry of a half-filled spherical tumbler rotated by less than 90° alternately about two perpendicular horizontal axes as a prototype to investigate mixing behaviors under competing influences of both stretching (during flow) and cutting & shuffling (upon changing the axis of rotation). From experiments utilizing x-ray subsurface visualization, we identify persistent non-mixing regions, whose properties are surprisingly consistent with predictions of a continuum model simulation and an idealized model based on the mathematics of piecewise isometries in which an object is cut, rearranged, and then reassembled into the original shape. Combining experiment, mathematical modelling and simulation, we aim to resolve the formation of non-mixing regions, and ultimately to predict non-mixing regions for a given protocol. The mechanisms of mixing and non-mixing behaviors of granular materials in this simple 3D geometry provide significant insights for mixer design in more complicated geometries. Funded by NSF Grants CMMI-1000469 and CMMI-1435065.

# 469635 Comparative Studies of Granular Flows Using Edem , Ligghts (open source) and PIV Experimental Setup

**Abstract:**

Granular flows and solid particle mixing are crucial to understand as they play a vital role in multiple disciplines, for example, from pharmaceutical, mining, chemical and nuclear engineering to astronomical applications like study of the formation and movement of sand dunes on Mars and other planets. This makes granular flows a really interesting and emerging field.

In this presentation, new results will be presented using open source software LIGGGHTS and compared with some of the previous results [1-2] using EDEM$^{TM}$. An experimental setup for PIV studies was also installed at Department of Chemical Engineering (DChE) in PIEAS and some of its results will also be presented.

**References**

[1] M.S. Siraj, S. Radl, B.J. Glasser, J.G. Khinast, Effect of Blade Angle and Particle Size on Powder Mixing Performance in a Rectangular Box. Powder Technology 211 (2011) 100-113.

[2] M.S. Siraj, Single-Blade Convective Powder Mixing: Impact of Blade Shape and Angle, Powder Technology 267 (2014) 289-301.

# 474892 Turbulent Mixing - From Fundamentals to CFD Models

**Abstract:**

Turbulent mixing is at the heart of numerous industrial processes. Over the past sixty-plus years significant advances in experimental and computational science have led to a fundamental understanding of how turbulent flows enhance mixing across a wide range of length and time scales. Despite the apparent complexity, this knowledge has led to predictive turbulent mixing models that can be incorporated in CFD simulations. In this talk, I will discuss how direct numerical simulations of turbulent mixing have been used to improve CFD models for turbulent reacting flows. I will also discuss open problems such as mixing models for differential diffusion, active scalars and multiphase flows.

# 474888 A Direct Numerical Simulation of a Turbulent Stirred Vessel Driven By a Rushton Turbine

**Abstract:**

New results will be presented from a Direct Numerical Simulation (DNS) of the just turbulent flow field, at Re=7300, in a stirred vessel provided with baffles and driven by a Rushton turbine. The simulations have been carried out by means of a Lattice Boltzmann solver, written in C, on a lattice comprising some 30 billion nodes, carrried out at the Irish Centre for High-End Computing ICHEC in Dublin.

The results of this simulation are compared with experimental data from the literature and with the results of the less refined DNSs reported earlier. The main variables of interest are the turbulent kinetic energy and its dissipation rate, phase averaged as well as phase resolved, which can be calculated from the outcome of such a DNS at any position in the vessel. Also turbulent spectra can be obtained in this way and some spectra will be presented for some positions in the vessel.

The idea behind performing DNSs for stirred vessel is to provide a yardstick for assessing the findings of Large Eddy Simulations and for evaluating the performance of sub-grid scale models essential for such LESs. Even experimental data may be assessed by means of a DNS as the resolution attainable in DNSs is better than in experiments.

# 449915 Analyzing the Closed-Loop Response of a Mixing Process Using Computational Fluid Dynamics and *Z*-Transform

**Abstract:**

Computational fluid dynamics (CFD) is widely used for analyzing the performance of mixing unit operations in the chemical process industry and can yield insights that are difficult or impossible to obtain by experimental studies. Motivated by the need to reduce the cost of repeated CFD simulations, we have previously demonstrated how to combine CFD with the $z$-transform technique to provide a convenient method for analyzing the open-loop responses of both a lumped-parameter mixing process (Yin and Yu, 2014 AIChE Annual Meeting Paper 354913) and a distributed-parameter mixing process (Yin, 2015 AIChE Annual Meeting Paper 404589).

For our current study, we consider a more complex problem of the closed-loop response of a mixing process, such as the temperature or concentration in a stirred-tank reactor with negative feedback control subjected to a load change. In the pure-CFD approach, one would implement the feedback loop and control action on the boundary conditions through the compilation and linking of user-defined subroutines. In the course of generating responses to study the mixing process and to optimize the control strategy, this procedure would have to be repeated for multiple CFD simulation runs.

Instead of taking the pure-CFD approach, a more efficient method is to combine CFD with the $z$-transform technique. After using CFD to obtain an open-loop response of the mixing process, we can derive the $z$-transform transfer function of the mixing process and then construct the $z$-space model of the closed-loop system. To generate additional responses of the closed-loop system would then require only fast polynomial arithmetic operations instead of full CFD simulation runs. This approach of combining CFD and $z$-transform not only reduces the computation cost, it also simplifies the investigation of the effects of various parameters on the response and stability of the closed-loop system.

# 473337 Modelling of RTD in a Complex Geometry Using Large-Eddy Simulation

**Abstract:**

Residence Time Distributions (RTD) can be used to analyze mixing in continuous systems, either stirred tanks or pipe lines. The RTD is determined by measuring the concentration of a tracer in the effluent from the system being studied in response to a perturbation in the concentration at the inlet (either a pulse or step change in the tracer concentration).

In this presentation we will demonstrate how a computational fluid dynamics (CFD) large-eddy simulation (LES) has been used to characterize the RTD in a real system with a non-standard geometry and this is compared to theoretical RTDs calculated analytically.

The computed RTD shows that the vessel can be modelled successfully using the tanks-in-series approach but also determines how many tanks are required.

Modelling of the RTD with LES CFD is substantially cheaper than making measurements with the real system and this work shows that the results can be used with confidence to assess the performance of a mixed system.

# 465158 Towards the Computational Optimization of Processing Equipment: Application to Static Mixers

**Abstract:**

The engineering performance of static mixer systems across a range of Reynolds numbers is predicted using time-accurate three-dimensional DNS and LES computational fluid dynamics (CFD) simulations. These simulations are executed using a lattice-Boltzmann algorithm over a voxelized solid-domain. Since this approach requires no user-meshing and minimal parameter specification, it is ideally suited for computational optimization. We begin by discussing the underlying physics governing the algorithm, and the difference between lattice-Boltzmann and conventional CFD. Next, for the Kenics UltraTab mixers operating in a turbulent regime we validate the predicted pressure drop and examine the transient eddy properties in the immediate vicinity of the mixer. We then validate the predicted mixing cup coefficient of variation (COV) against the experimentally measured values. Next, for laminar systems, DNS simulations are performed on Kenics KM mixers to predict the pressure drop and downstream mixing properties. The predicted pressure drop is compared to experimental measurements and literature values. The predicted COV is also validated against experimental data. We conclude by discussing how this modeling approach can be combined with computational search heuristics to identify optimized static mixer topologies.

# 469089 Mixing and Age Distribution in Turbulent Jet Mixers

**Abstract:**

In this talk, mixing and age distribution in industrial turbulent jet mixers will be discussed. Spatial distributions of mean age and higher moments of age in the reactors are obtained from CFD solutions of the steady transport equations. The spatial distribution of mean age reveals details of the spatial non-uniformity in mixing. Variances of age and the degree of mixing discussed by Danckwerts and Zwietering are computed. Blend time can be estimated from the mean age probability density function curves. Two types of jet mixers, angled jet and vertical jet mixers will be considered. Both types are widely used in process industries. Mixing and age distribution in the angled jet mixers have been presented previously and the main results will be reviewed. The results showed that the predicted blend time using mean age method agrees well with lab experiments and existing correlations. On the other hand, there has been no blend time correlations for vertical down flow jet mixer. Mean age method will be used to predict blend time and the results will be compared with available experimental data in literature. A correlation will be proposed based on the predicted blend time.

# 471017 Mean Age Theory for Mixing Time: Experiments and Mesh Type Effects on CFD Simulation

**Abstract:**

The optimal meshing and simulation strategy for efficiently simulating mixing time problems can cut down on overall simulation time and increase engineering output. Steady state mean age theory and traditional transient simulation have been applied towards measuring mixing time in stirred tank vessels in order to study the impact of various meshing types. Polyhedral, tetrahedral and hexahedral meshes were simulated using STAR-CCM+. Simulations were conducted across a range of varying impeller speeds and configurations. Accuracy to experimental data and computational time were key response variables in the study. Very close matching was obtained between the steady state mean age method, the traditional transient CFD method, and experimental data, which indicates the reliability of mean age to study these problems. Additionally, some data regarding calibrating the mean age mixing type method to other mixing time measurements will be presented.

# 461380 Viscous Solid-Liquid Mixing in Agitated Vessels Via CFD-DEM Simulation: Suspension Dynamics and Influence of the Geometry

**Abstract:**

Solid-liquid mixing in mechanically agitated vessels is a unit operation found in many industries such as ore processing, pharmaceutical and cosmetics. The majority of the efforts concerning the study of solid-liquid mixing has been geared toward the assessment of the minimum agitation speed (or just-suspended speed) $N_{js}$ required to suspend the particles in the turbulent regime. Moreover, Zwieteringâ€™s correlation is known to perform poorly in the laminar and transitional regimes [1, 2]. In particular, experiments by Lassaigne *et al.* [1] revealed that the hydrodynamics of non-dilute suspensions depends on the flow regime, and the influence of the fluid viscosity and particle diameter is in opposition to what is predicted by Zwieteringâ€™s correlation. The fact that it can be challenging to investigate experimentally the behavior of non-dilute suspensions, which usually are opaque systems, has provided the impetus for the development of simulation models.

A variety of models have been developed to simulate solid-liquid flows. These include the classical Eulerian-Eulerian (or two-fluid) model, and the combination of the Discrete Element Method (DEM) for the particles and CFD methods for the liquid phase. Recent work in our group has shown that such a Eulerian-Lagrangian CFD-DEM approach can be highly accurate in predicting the fraction of suspended solids and the flow patterns, when compared to experiments [3]. However, the potential of this model has yet to be exploited to improve our understanding of solid-liquid dynamics and help design and optimize such mixing operations.

In this work, an unresolved CFD-DEM model developed by our group, which is based on the CFDEM framework combining LIGGGHTS for the DEM and OpenFOAM for the CFD, is used to study the solid-liquid mixing of non-dilute suspensions stirred by a pitched blade turbine (D=T/3). The flow patterns and solids concentrations are investigated for three impeller clearances (T/3, T/4, T/5), with and without baffles. The fraction of suspended solids

and the degree of mixing are assessed from the particle trajectories and the so-called Lagrangian suspended fraction analysis technique.

**References**

[1] Lassaigne M., B. Blais, L. Fradette, and F. Bertrand, Experimental investigation of the mixing of viscous liquids and non-dilute concentrations of particles in a stirred tank, Chemical Engineering Research and Design, in press.

[2] Ibrahiman S. and A. W. Nienow, Comparing impeller performance for solid-suspension in the transitional flow regime with Newtonian fluids, Chemical Engineering Research and Design, 77, 721-727 (1999).

[3] Blais B., M. Lassaigne, C. Goniva, L. Fradette and F. Bertrand, Development of an unresolved CFD-DEM model for the flows of viscous suspensions and its application to solid-liquid mixing, Journal of Computational Physics, in press.

# A fully coupled CFD-DEM algorithm for S-L mixing in stirred vessels

*Thomas Eppinger*
*Oleh Baran*
*Ravindra Aglave*
*Simon Lo*

Material presented at AIChE 2016 meeting in San-Francisco on November 15, 2016

## Abstract

Solid-Liquid mixing in high viscosity liquid and/or high solids concentration is a challenging problem due to the solid-solid and solid-liquid phase interactions. Just suspended speed ($N_{js}$), suspension height ($H_s$) and the power consumption can vary not only with the material properties but also with the design of tank and impellers.

In the past, there has been work presented using multiphase CFD methods, specifically the Eulerian approach. Such approach has shown promising results for turbulent conditions (Aglave, et al., 2016). Further work has also been carried out where Discrete Element Modeling for particle flows was co-simulated with CFD simulations for fluid flow (Bertrand, et al., 2016). Often, this could mean creating and maintaining two different codes for the two portions of the flow.

In this work we would be presenting a method of performing such simulations in a coupled way using a single code STAR-CCM+ rather than a co-simulation framework. For DEM phase "pressure gradient" and "drag forces" are included in integrating equations of motions for each particle. The continuum field for fluid solution is obtained from solving Navier-Stokes equation based on the local mean variables over computational cells. These local variables include the solid fraction for the sources of momentum due to particles. Comparisons with existing literature data will be presented which represents a lab operation of the stirred vessel. Future outlook on translating this for a more practical operation of the vessel will be outlined.

## Introduction

In this work we simulate the mixing experiments reported in (Lassaigne, et al., 2016). In our approach we do not use computationally very expensive immersed boundary method (Blais, et al., 2016). Instead, the CFD solver has an access to the current concentration of particles in each cell and uses it to impose momentum sources in fluid to enforce Newton's third law in momentum exchange between particles and fluid. To achieve acceptable accuracy we introduce DEM-CFD coupling which uses original fine mesh for fluid phase and coarser mesh for DEM-CFD coupling. Commercial engineering software package STAR-CCM+ version 11.06 was used to run DEM coupled to CFD models to produce results discussed in this paper.

## Simulated system

The simulated geometry is shown in Figure 1. The cylindrical container (tank) of mixing vessel has diameter $T = 365\ mm$. While in experiment the liquid fill level is lower than the height of the container, in simulations we adopted an approximation of setting the height of the tank to be the same as the height of the liquid level in experiment so we don't include into the model neither air nor the free surface between liquid and air.

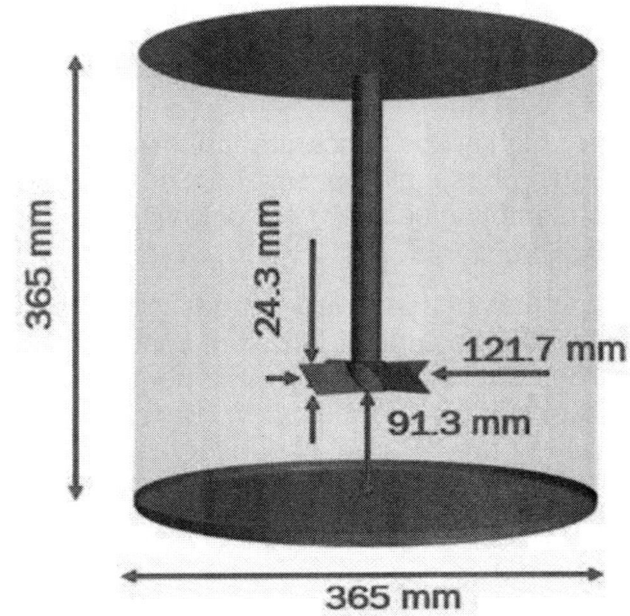


| Tank diameter | T=0.365 m |
|---|---|
| Liquid level in tank | H=T |
| Impeller type | Pitched-blade turbine |
| Impeller Diameter | D=T/3 |
| Off-bottom clearance | C=T/4 |
| Blade width | $W_i$=D/5 |

**Figure 1: Stir vessel geometry dimensions (left), and dimension details (right) using notations from (Lassaigne, et al., 2016)**

A pitch-blade turbine impeller of diameter $D = \frac{T}{3} = 121.7\ mm$, pitch angle $\alpha = 45^o$, blade thickness $e = \frac{D}{100} = 1.2\ mm$, vertical component of blade width $W_i = \frac{D}{5} = 24.3\ mm$ (which corresponds to actual blade width being $\frac{W_i}{\cos\alpha} - 2e = 32\ mm$), and shaft diameter $D_{shaft} = \frac{D}{5} = 24.3\ mm$ is placed above the bottom of container with off-bottom clearance $C = \frac{T}{4} = 91.3\ mm$. The tank is completely filled with glucose solution of density $\rho_f = 1400\ kg/m^3$ and dynamics viscosity $\mu = 1\ Pa\ s$. 150,000 spherical glass beads (modeled by DEM) are settled by gravity at the bottom of the tank. DEM model parameters reflecting material properties of particles and geometry are summarized in Table 1.

**Table 1: Material and contact properties of particles and geometry**

| | Material | E (MPa) | Poisson ratio | Density (kg/m$^3$) |
|---|---|---|---|---|
| Particles | glass beads | 100 | 0.35 | 2500 |
| Geometry | Aluminum | 68000 | 0.33 | 2702 |

| | Friction | Restitution |
|---|---|---|
| Particle-Particle | 0.6 | 0.1 |
| Particle-Geometry | 0.1 | 0.1 |

Impeller rotates in clockwise direction to pump liquid towards the bottom of tank (see Fig 2. Left). Eleven different rotational velocities of impeller in the range between 30 and 550 rpm were simulated.

## Numerical method for coupling between solid and liquid flows

The DEM modeling and DEM coupling to CFD approach used in this work are extensions of the method described in (Baran, et al., 2013). In particular, for the DEM-CFD coupling we used novel algorithm for source smoothing with cell clustering (Baran, et al., 2016) described in more details below.

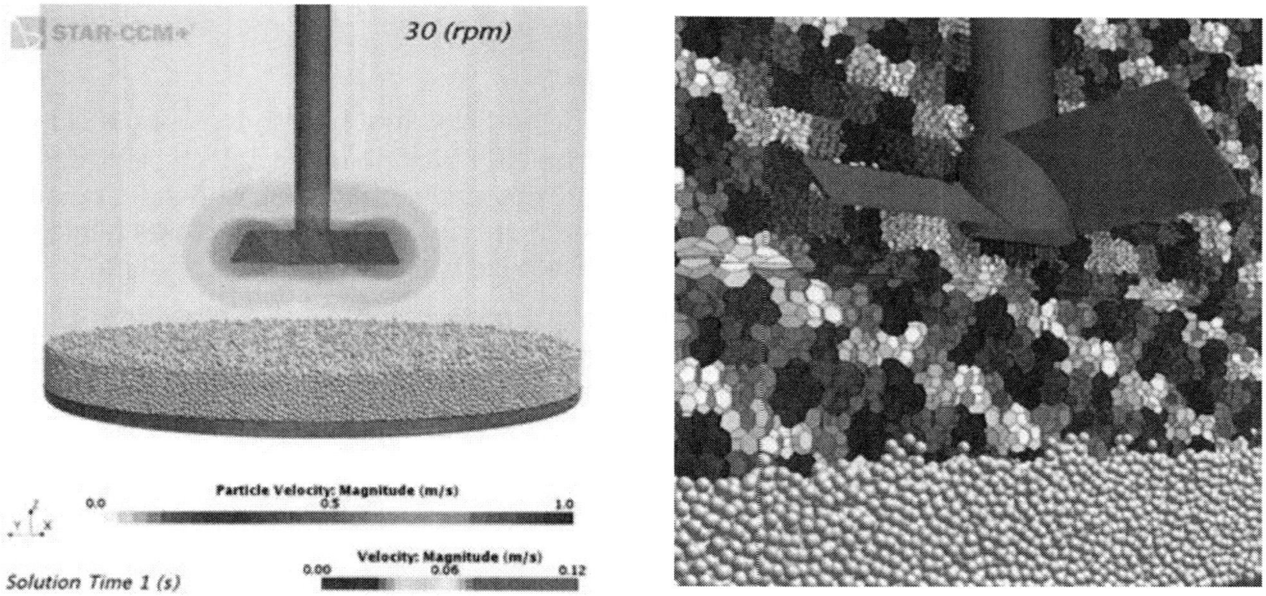


**Figure 2: Left: Steady state configuration at impeller velocity 30 rpm with particles being submerged in liquid and resting at the bottom. Particles colored by velocity using white-green color scheme; Fluid velocity is shown on vertical plane cross-section using blue-red-highpass color scheme (with opacity zero for low fluid velocities). Right: Zoom into volume cells of polyhedral mesh around the impeller (clipped view) with cells colored by association with cell cluster.**

Eulerian liquid phase solution is obtained by solving the following finite volume discretized Navier-Stokes equation (2) in a way consistent with Hydrodynamic Model A formulation (Bouillard, et al., 1989)

$$\frac{\partial(\epsilon\rho_f)}{\partial t} + \nabla\cdot(\epsilon\vec{u}) = 0 \tag{1}$$

$$\frac{\partial(\epsilon\rho_f\vec{u})}{\partial t} + \nabla\cdot(\epsilon\rho_f\vec{u}\vec{u}) = -\epsilon\nabla \mathrm{p_w} + (1-\epsilon)\beta(\langle\vec{v_p}\rangle - \vec{u}) \tag{2}$$

$$\rho_p\Omega_p\frac{d\vec{v_p}}{dt} = -\Omega_p\nabla p_s + \Omega_p\rho_p\vec{g} + \Omega_p\beta(\vec{u}-\vec{v_p}) + \sum_j \vec{F_j} \tag{3}$$

Here working pressure $p_w$is the difference between static pressure $p_s$ and hydrostatic pressure $\rho_f gh$, $\vec{u}$ is a velocity of fluid in a cell, $\vec{v_p}$ is a velocity of particle, $\langle\vec{v_p}\rangle = \sum(\Omega_p\vec{v_p})/\sum\Omega_p$ is a volume averaged particle velocity in a cell, $\Omega_p$ is the volume of particle, $\sum_j \vec{F_j}$ is the sum of all contact forces due to collisions with neighboring particles, $\beta$ is a friction coefficient which is set to reproduce Ergun equation based drag force for particles in cells having higher solid fraction and Wen-Yu drag force for the particles in cells with lower solid concentration. The void fraction $\epsilon$ used in above equations is defined on coarser grid generated by clustering original cells of Eulerian grid into the conformal clusters of cells of roughly similar cluster volume (see Fig 2 right)

$$\epsilon = 1 - \frac{\sum_{cluster}\Omega_p}{\Omega_{cluster}} \tag{4}$$

Here $\sum_{cluster}\Omega_p$ is the sum of the volume of all particles that belong to a single cluster of cells of volume $\Omega_{cluster}$. The process of calculating the void fraction using cell clusters can provide smooth and still accurate information on particle concentrations to fluid solver even for the Eulerian grids with high gradient of cell sizes and does not require the particles being smaller than the size of smallest cell (as would be the case without source smoothing).

In our case of stirred vessel we first generated a grid with total 517,070 volume cells and then grouped all the cells into 43,357 clusters of approximately equal volume, see Fig 2 right. The average size cluster can fit on average about 35 glass beads.

In our model we set K-Omega turbulence model for the liquid phase and neglected possible contributions from shear lift force on particle, spin lift force on particle, virtual mass.

## Simulation details and results

To approach the steady state we followed the following steps. First, 30 rpm case was set and simulated till final physical time 10 seconds. Then we started ramping up rpm linearly every second to the next target value of rpm and at the end of each ramping second starting new separate simulation with constant target rpm, see Fig. 3 below.

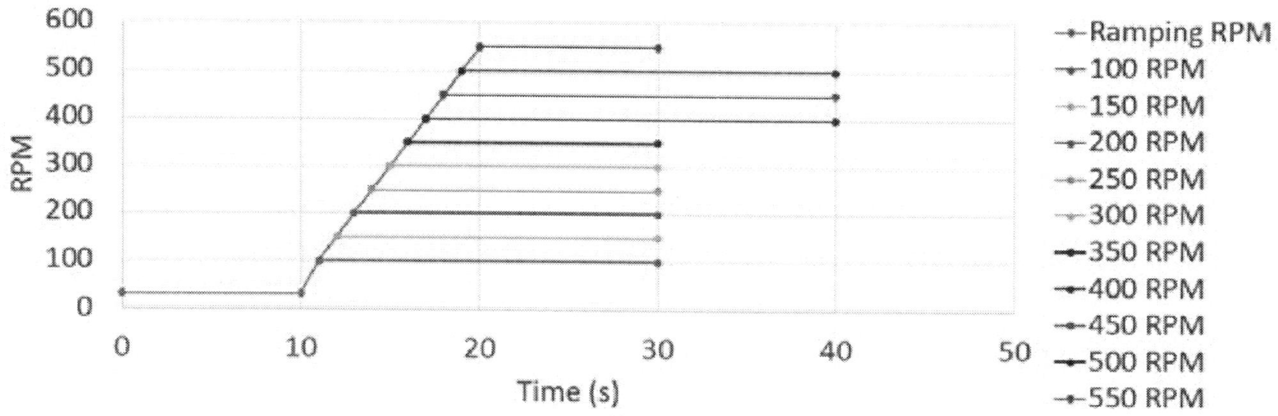


**Figure 3: Achieving steady state through ramping rpm to target value. Length of horizontal lines indicates the number of seconds simulated at constant rpm for each simulated case**

For example, to obtain final steady state at 400 rpm we first simulated 10 seconds at 30 rpm, then linearly ramped it from 30 rpm to 100 rpm for $10\,s < t < 11\,s$, then linearly ramped impeller's velocity from 100 rpm t0 400 rpm for $11\,s < t < 17\,s$, and then ran at 400 rpm for 23 seconds until $t = 40\,s$. This way we reduced the risk of shocks in the system and numerical instabilities in model that could have been caused by sudden changes in impellers' velocity.

Table 3 below shows the snapshots of particle and fluid states taken at the final time for each impeller velocity.

**Table 3: Simulated impeller velocities and corresponding states of particles and liquid in steady or close to steady state**

| 30 rpm | 100 rpm | 150 rpm | 200 rpm |
|---|---|---|---|

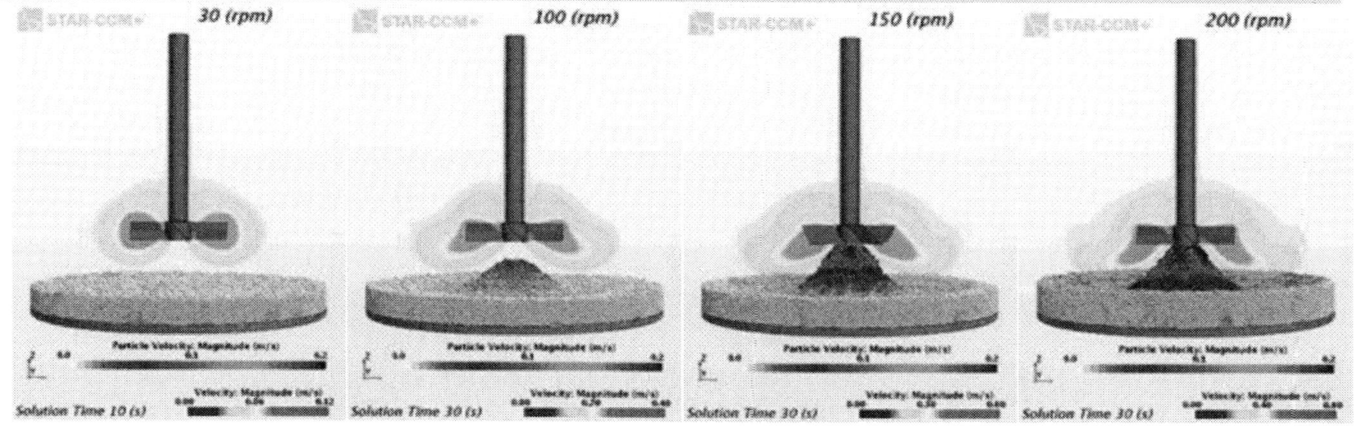

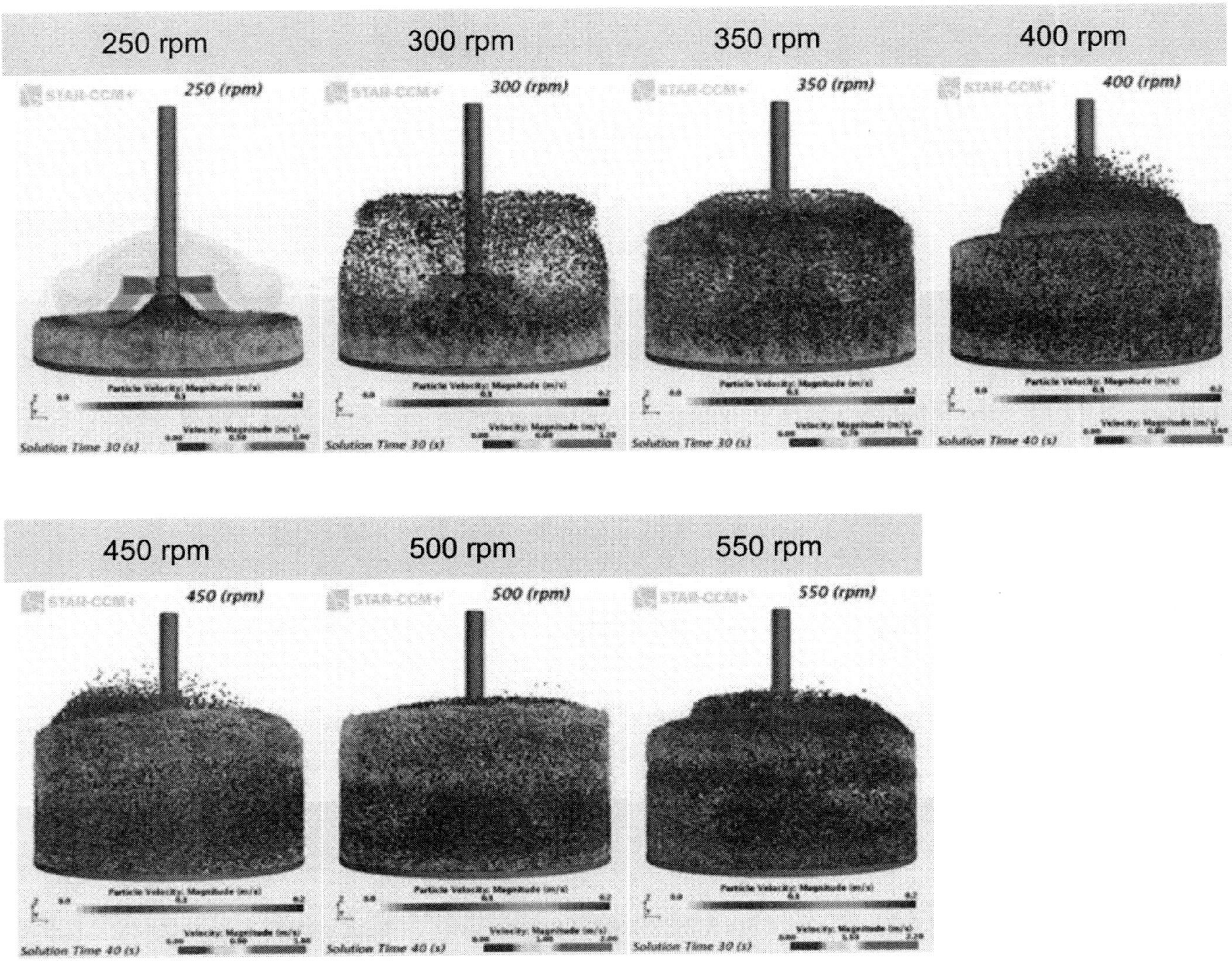


Above dependence of mixture state on the impeller speed agrees well with experimental data reported in (Lassaigne, et al., 2016).

In simulations we monitor the number of suspended particles using two methods. First method tracks the compression force on each particle due to contacts with other particles or geometry and puts the particle into suspended particles category if the stress is zero and into settled particle category otherwise. In the second method we use "connected components" analysis to differentiate the suspended particle from settled. Within this method the particle's "connected component size" is evaluated for each particle, see Fig. 4 below.

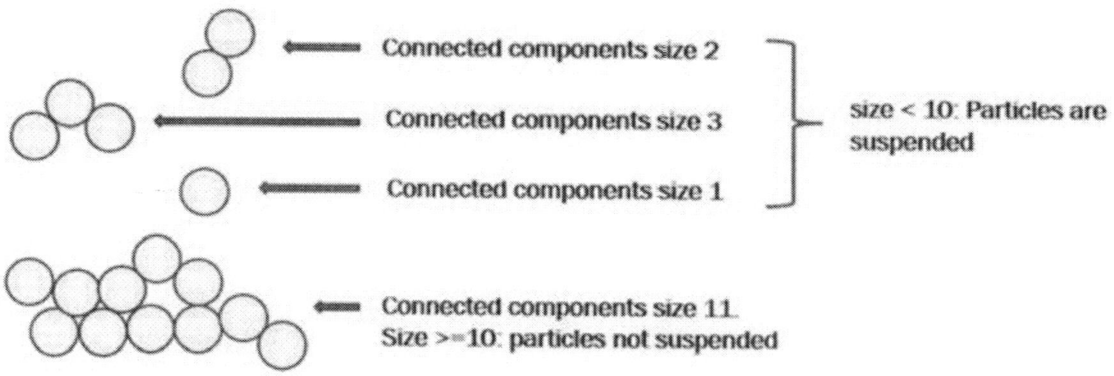


**Figure 4: Schematics illustrating the meaning of "connected component size"**

If the connected component size is smaller than 10 we put the particle into settled category and the rest of particles go into the suspended particles category. Fig. 5 below shows the results for the number of suspended particles (averaged over last 2 seconds) as a function of impeller velocity, solid red curve showing the results with using connected components criteria and red dashed curve showing the results with using compression force criteria. For the reference we plot with black curve the number of suspended particles data obtained in experiment using pressure gauge technique (Lassaigne, et al., 2016)

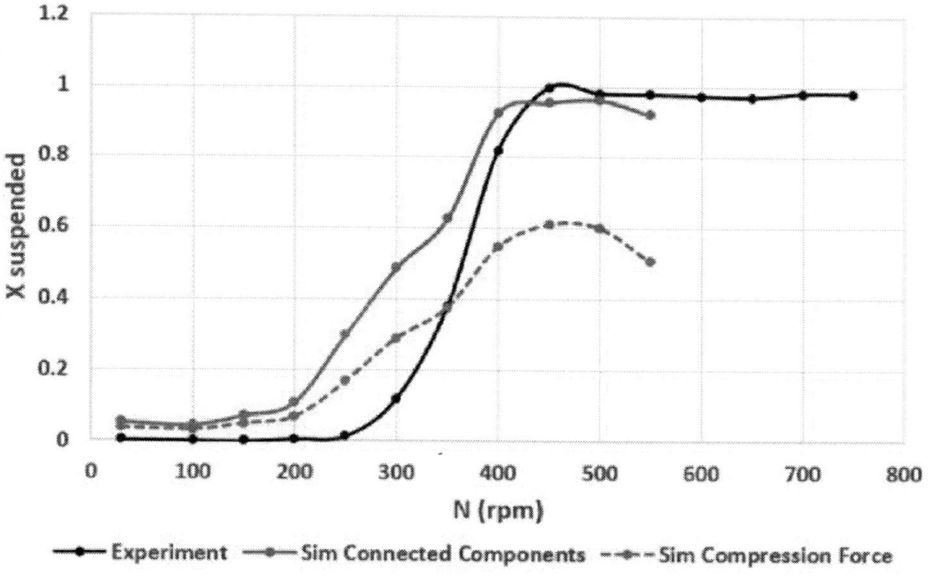


**Figure 5: Results for the number of suspended particles directly counted in the simulations (red curves) compared to the number of suspended particles extracted from experiments (Lassaigne, et al., 2016) using pressure gauge technique (black curve)**

The connected component criteria provides better match with experiment compared to the compression force criteria. Assuming the number of suspended particles is indicative of the "fluidization" state of the mixture, we can conclude that the fluidization of particles due to liquid flow caused by impeller motion happens at slightly lower rpm in simulations compared to experiment which could be attributed to the particular choice of the threshold value 10 for the connected component size used to differentiate suspended particles from settled.

Finally, in simulations we measure the surface average working pressure (difference between static and hydrostatic pressure) at the bottom of tank and compare it with the results for the pressure measurements at the bottom in the experiments, see Fig 6 below

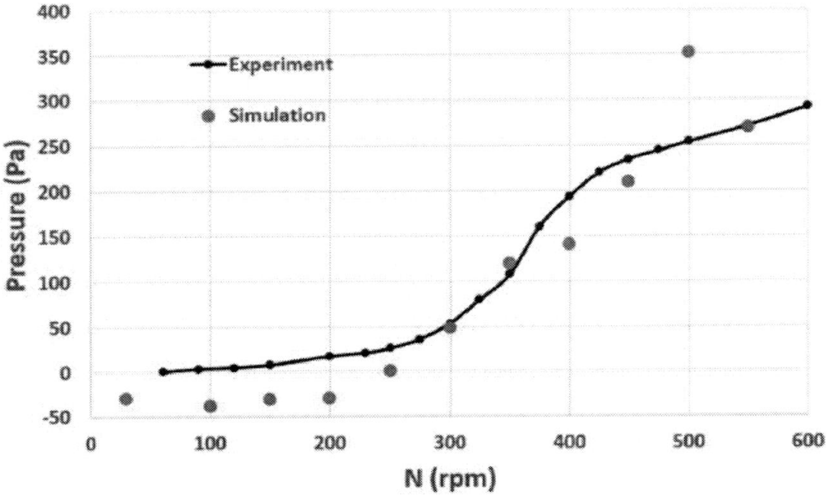


**Figure 6: Results for the surface average pressure at the bottom (red dots) compared with pressure measured by pressure probe at the bottom in experiment (black curve)**

The results for pressure in simulations show the increase of pressure due to fluidization happening at the same range of impeller velocities as in experiments, thus validating our model.

## Summary

We introduced 2-way coupling DEM-CFD model for stirring solid particles in glucose solution which uses novel "Source Smoothing with Cell Clustering" coupling algorithm to attain the acceptable for this application balance between accuracy and performance. The typical simulation time is less than 2 hours to simulate 1 second of physical time on 24 cores. Using our model we reproduced fluidization of particles happening at about the same range of rpms as observed in experiments. We also introduced new "connected components" method of counting suspended particles in DEM simulations.

## References


**Aglave Ravindra [et al.]** Predicting power consumption at high solids concentration in mixing vessels. - [s.l.] : Internal Report, 2016.

**Baran Oleh, Greening Doran and Kodl Petr** Numerical Simulations of Fluidized Beds in Fine-Grid Geometries Using Source Smoothing with Cell Clustering Method for DEM-CFD Coupling [Conference]. - San-Francisco : AIChE, 2016.

**Baran Oleh, Kodl Petr and Aglave Ravindra** DEM Simulations of Fluidized Bed Using Scaled Particle Approach [Conference] // AIChE. - 2013.

**Bertrand Francois [et al.]** A robust and efficient CFD-DEM method fot investigation of viscous solid-liquid mixing. - 2016.

**Blais Bruno [et al.]** A semi-implicit immersed boundary method and its application to viscous mixing [Journal] // Computers and Chemical Engineering. - 2016. - Vol. 85. - pp. 136-146.

**Bouillard J. X., Lyczkowski R. W. and Gidaspow D.** Porosity Distributions in a Fluidized Bed with an Immersed Obstacle [Journal] // AIChE Journal. - 1989. - 6 : Vol. 35. - pp. 908-921.

**Lassaigne Manon [et al.]** Experimental investigation of the mixing of viscous liquids and non-dilute concentrations of particles in a stirred tank [Journal] // Chemical Engineering Research and Design. - 2016. - Vol. 108. - pp. 55-88.

# 465606 "Validation of the Coalescence-Dispersion Model of Mixing for Complex Chemical Reactions"

**Abstract:**

The integration of the coalescence-dispersion (c-d) model of mixing into a CFD simulation code has been extensively tested. The c-d model simulates mixing by assuming that the fluids may be represented by mobile fluid points which carry the mass, concentrations, and other properties of the fluids. Upon coalescence of two randomly chosen fluid points, their properties are averaged (concentrations, enthalpies, etc.). Upon dispersion (separation) the two fluid points have identical properties. In a c-d simulation of a chemical reaction, the reactions occur for the length of the time increment between flow events as if the fluid points are batch reactors. Use of the c-d model avoids the need for a closure model for the mixing in the presence of chemical reactions, which has only been successful for two competing reactions.

Integration of the c-d model into a CFD simulation requires that the mobile fluid points move, mix and disperse in a way dictated by the CFD simulation. Each volume segment of the CFD simulation is assigned a number of fluid points which move from segment to segment as dictated by the CFD simulation. This is accomplished by a subroutine linked to the CFD code that accesses the velocity and mixing rate predictions of the CFD. A critical aspect of the c-d model is in the most effective choices of the time increment and the number of fluid points in each CFD volume segment. This is dependent on the number of c-d events during each time increment and the flow volumes from segment to segment.

The validations have been done by comparing results with well-established experimental data for single and multiple reactions in various types and sizes of chemical reactors. The validations show that the c-d simulation method can be a valuable addition to experimental research on multiple competitive reactions where there is interest in yields of given products.

# 460117 An Open-Source Adaptive-Mesh Finite-Element Framework for the Modelling of Polydispersed Flows Using Population Balances

**Abstract:**

Polydispersed multiphase flows are an essential part of the chemical industry, found in applications such as minerals processing, food processing, and pharmaceutical, to name a few. Computational fluid dynamics (CFD) has become a popular tool for analysing the hydrodynamics in industrial devices (such as bubble columns and agitated mixers) in the past decade due to an improvement in the computational infrastructure and the numerical methods to model multiphase flows. However, there are certain limitations of the current modelling frameworks that prevent them from tractably simulating industrial-scale processes. The multi-scale nature of the systems leads to inter-particle interactions creating polydispersity, which is sometimes neglected in CFD models due to its expensive nature for the modelling of industrial-scale systems. This work focusses on the development of an efficient modelling framework that allows for the simulation of the polydispersed multiphase flow hydrodynamics in Fluidity, an open-source finite-element code, using adaptive meshes. The direct quadrature method of moments (DQMOM) was implemented in Fluidity as part of this work to solve the population balance equation (PBE). Rigorous verification and benchmarking is presented, followed by model validation on gravity-driven turbulent flow in bubble columns. The framework is then applied for the modelling of a flotation column, which contains two polydispersed phasesâ€"gas bubbles and solid mineral particlesâ€"in a liquid continuum. The attachment and detachment of hydrophobic mineral particles to the gas bubbles is also modelled using appropriate expressions. The above simulations are performed using fully-unstructured anisotropic adaptive meshes, which clearly show an improvement in the efficiency of the solution method. The benefit of modelling the polydispersity of gas and solid phases is also demonstrated in this work for flotation columns.

# 461791 Modeling and Simulation of Flash Nanoprecipitation in a Multi-Inlet Vortex Reactor

**Abstract:**

Flash nanoprecipitation (FNP) is an emerging micellization technique applicable to making functional nanoparticles for drug delivery, drug targeting, bio-imaging, sunscreen, and food technology. FNP utilizes rapid mixing of an anti-solvent stream and a solvent stream containing stabilizing block copolymer and nanoparticles. The mixing of these two streams induces supersaturation, leading to the formation of functional nanoparticles. One constraint of FNP is to have a fast mixing time compared to nanoparticle formation time, which lead to the design and scale-up study of the multi-inlet vortex reactor (MIVR). Experiments have shown promising results for tunable, consistent nanoparticle sizes, but in order to optimize the MIVRâ€™s design, the mixing behavior of the reactor needs to be examined. Computational fluid dynamics (CFD) simulations have been used as an aid to benchmarking the mixing and determining the flow patterns of the MIVR. In order to model FNP in the MIVR correctly, the turbulence field inside the MIVR has to be predicted correctly first. Considering the complex turbulent swirling flow inside the MIVR, our recent work has applied a hybrid RANS/LES method for turbulence modeling successfully. Population Balance Equations (PBE) can be coupled with CFD simulations to describe particle size evolution over time, but this direct coupling is computationally intractable. As an alternative, Quadrature Based Moment Methods (QBMM) are used to compress computational costs by tracking the moments of the particle size distributions. The Conditional Quadrature Method of Moments (CQMOM) is used to describe the aggregation of block copolymers and nanoparticles. As a means to streamlining the simulation process, OpenQBMM, an extension of OpenFOAM, is used to solve the moment transport equations along with appropriate mixing and aggregation kernels. By utilizing recently developed and improved CFD models and QBMM, this work will increase the understanding of FNP in the MIVR.

# 472840 Characterizing Flow and Solid Suspension in Optimax Crystallization Workstation

**Abstract:**

**Characterizing Flow and Solid Suspension in OPTIMAX Crystallization Workstation**

M. Sardeshpande[1], A. Pandit[1], S. Vedantam[2] and V.V. Ranade[1*] [1]Industrial Flow modeling Group Chemical Engineering and Process Development Division National Chemical Laboratory, Pune 411008, INDIA [2]Chemical Engineering Division, Indian Institute of Chemical Technology, Hyderabad – 500007, INDIA *Email: vv.ranade@ncl.res.in

OPTIMAX® (of Mettler-Toledo Ltd.) is one of the widely used crystallization workstations. OPTIMAX® is an automated reactor, offering heating and cooling electrically on the basis of Peltier technology. It is usually equipped with Focused Beam Reflectance Measurement (FBRM) probe for measuring on-line Chord Length Distributions (CLD). A schematic of the workstation is shown in Figure 1. In order to appropriately interpret the data obtained from OPTIMAX® workstation, it is essential to quantitatively understand flow, heat transfer and solid suspension in it. In this work we have developed a CFD model to characterize flow and solid suspension in OPTIMAX® workstation.

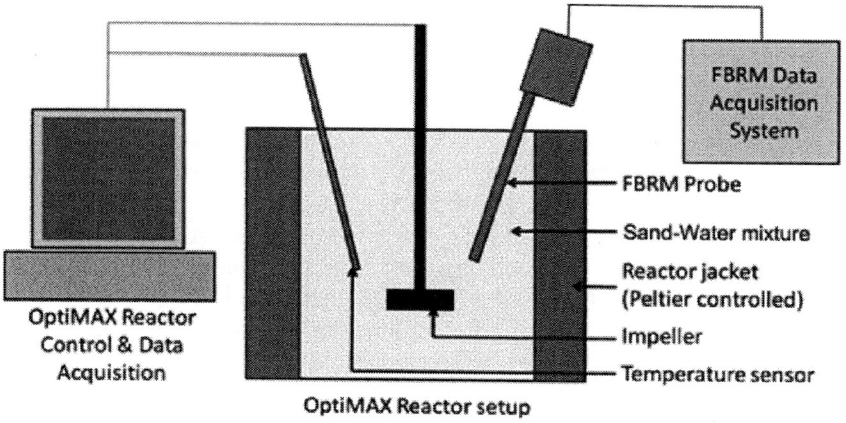


Figure 1: Schematic of OPTIMAX® crystallization workstation

In order to characterize solid suspension in OPTIMAX® workstation, we have used the approach proposed by Sardeshpande et al. (2010). Sardeshpande et al. (2010) reported hysteresis in cloud height observed experimentally in a stirred tank reactor while increasing and then decreasing the impeller speed. The hysteresis was then captured using a CFD model which served as a rigorous validation for the model. In the present work we wish to capture the hysteresis on increasing and then decreasing the impeller using the FBRM measured particle counts. A sample of the particle counts measured by the FBRM for increasing impeller speeds is shown in Figure 2. This relies on measurement of hysteresis in particle volume fraction (or number of particle counts) with respect to rotational speed of impeller. Experiments were carried out with baffled, hemi-spherical bottom stirred tank reactor. 1000ml of this OPTIMAX reactor contents Focused Beam Reflectance Measurement (FBRM) probe (used for estimating the Particle Size Distributions (PSD)). Hysteresis measurements were carried out for three different particle volume fractions using glass beads.

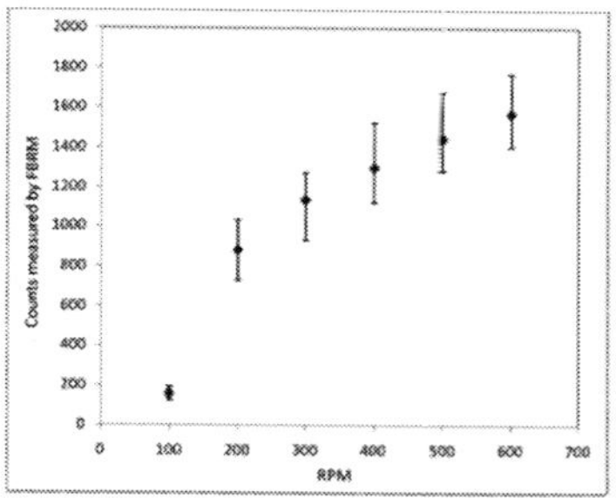


Figure 2: Particle Counts measured by FBRM as a function of impeller speed while increasing impeller speed

CFD model of the considered configuration was developed. Multiple reference frame approach was used to model the rotating impeller and multiple phases have been simulated using the Eulerian - Eulerian approach. Appropriate care was taken to ensure that simulated results are independent of numerical parameters (grid size, discritization scheme, time step and so on). The simulated results of hysteresis in particle counts are compared with the experimental data. Critical analysis of obtained results was carried out.

The presented approach, model and results will be useful to characterize flow and solid suspension in the OPTIMAX® workstation. The developed CFD model will be useful for eventual simulations of crystallization process occurring in OPTIMAX® workstation as also in crystallizer scale up studies.

References Sardeshpande M. V., Juvekar V. A. and Ranade V.V., 2010 Hysteresis in Cloud Heights During Solid Suspension in Stirred Tank Reactor: Experiments and CFD Simulations, AIChE Journal, 56(11), 2795-2804

# 451940 Limitations of Simple Mass Transfer Models in Polydisperse Liquid- Liquid Dispersions

**Abstract:**

In this work turbulent liquid-liquid dispersions in stirred tanks are simulated by using computational fluid dynamics (CFD), population balance models (PBM) and different simplified models. The evolution of the droplet distribution is evaluated by means of the PBM in all simulations; different strategies are instead adopted for the description of the flow field: zero-dimensional (0D) models, in which perfect mixing of the phases is assumed, and three-dimensional (3D) models, where the spatial inhomogeneities are considered through the numerical solution of mass and momentum balances. 3D simulations are performed with our implementation of the quadrature method of moments (QMOM) into the Eulerian-Eulerian two-fluid solver of the open-source CFD code OpenFOAM, while 0D models use different Quadrature-based moment methods (QBMM) for the solution of the PBM. Different coalescence and breakage kernels are considered, based on the classical homogeneous and multifractal turbulence theories. Model predictions are eventually validated against experimental data for different test cases, consisting of different stirred tank geometries and different continuous and disperse phases. The comparison of the different (0D and 3D) models clearly shows that the 0D models can be used under dilute conditions, whereas in the dense regime the 3D models should be used instead. The comparison allows also to derive a correct formulation of the 0D models, that are instead often wrongly formulated.

# 465487 Simulation of Liquid-Liquid Dispersions in Continuous-Flow Pump-Mixer with CFD-PBM

**Abstract:**

Liquid-liquid dispersions in continuous-flow pump-mixers are frequently used in the solvent extractions of rare earths. The extent of mixing is commonly judged by hold-up and drop size distribution, which are affected by the interaction of droplets seriously. In this work, droplet breakage and coalescence in turbulent flow are simulated via computational fluid dynamics (CFD) simulations combined with population balance model (PBM). The drop size distributions and hold-up of dispersed phase are predicted in different dispersions created in a continuous-flow mixer with a both sides shrouded turbine with radial trapezoidal blade(BSTRTB), with aqueous phase as continuous and organic phase as disperse phase. Effects of impeller speed, flow ratio and physical properties on hydrodynamics features are investigated. Simulations are performed with multiple reference frame (MRF) model and Eulerian-Eulerian two-phase mothed. Two population balance models, differing in droplet breakage kernels, have been evaluated. The experimental data of velocity field, obtained by using particle image velocimetry (PIV), and hold-up in system of kerosene/water has been used to compare the models for their ability to predict the mixing characteristics. The better one is applied to forecast the mixing features of the quasi real rare earth extraction systems.

# 468248 Investigating Bubble Size Distribution in Fermentation Reactors with CFD


**Abstract:**

Fermentation is a common method for producing bio-products such as alcohols and pharmaceutical compounds or alternative feedstocks. In the primary stage, presence of oxygen drives the metabolism and hence growth of the organic substance. Oxygen is supplied via air in stirred vessel type fermenters. While a certain configuration tested in a lab may result in satisfactory distribution of the dissolved oxygen (DO), large scale fermenters need careful design considerations.

Computational methods (CFD) have been reported in large amount of literature to investigate the flow patterns and consequently the gas hold up in such equipment. However, the air bubble injected vary in size throughout the reactors due to the choice of impellers, rpm, baffles, and other internals such as coils or dip tubes. The size of the air bubbles affects directly the gas hold up and the interfacial area. Therefore it is necessary to account for the varying bubble size distribution due to coalescence and breakup.

In this work, we present the use of a size distribution model to account for the bubble size distribution called S-gamma model in the commercial code STAR-CCM+. This model shows the differences in assuming constant bubble size but is not as computationally expensive as the quadrature (DQMOM/QMOM) type of models.

Features of the S-gamma model will be discussed and a case study presented where the two methods are compared. An outlook on future work and advances will also be given.


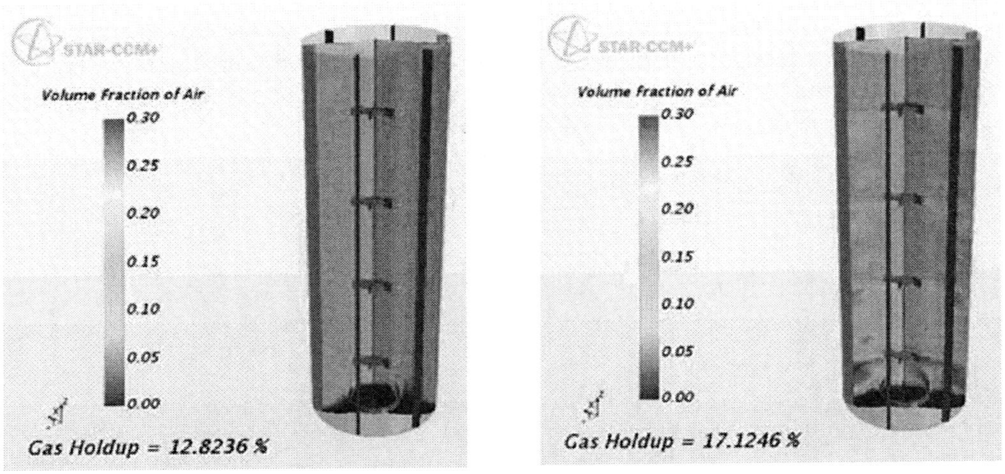


*Gas distribution and total gas hold-up for a plant scale bioreactor assuming a constant bubble size (left) and a bubble size distribution including coalescence and break-up (right).*

References:


Lo, S. and Zang, D., Modelling of Break-up and Coalescence in Bubbly Two-Phase Flows, Journal of Computational Multiphase Flows, Vol 1(1), p 23, 2009

# 469644 Experimental Measurement and CFD-PBM Simulation of Bubble Size Distribution in a Gas-Liquid Taylor Vortex Reactor

**Abstract:**

Bubbly gas-liquid Taylor-Couette vortex flow has gathered increasing interest because of its potential applications to a variety of chemical and biochemical processing problems. The evolution of the bubble size distribution in these systems influences the hydrodynamics and interphase mass transfer. In order to quantitatively capture the bubble size evolution and its effect on the hydrodynamics, a two-fluid CFD-PBM model was developed for the simulation of bubble-bubble interactions (e.g. coalescence and breakage) in bubbly turbulent Taylor-Couette flow and compared with experimental data. The direct quadrature method of moments (DQMOM) was used to solve the populous balance model (PBM) as it requires relatively low computational cost. The performance of several widely used combinations of aggregation and breakage kernels was assessed, and simulations were shown to accurately predict bubble size distribution and bubble volume fraction for a wide range of flow conditions. These results may provide important guidance for reactor design and scale-up.

*Non –Newtonian Fluid Dynamics in a Three Phase Bubble Column: Experimental and Computational Fluid Dynamics (CFD) Comparison*

PETROBRAS UNICAMP PQGe

Introduction | Experimental Unit | Results and discussion | Conclusions

48 YEARS
41.000 STUDENTS
70 CURSOS
24 INSTITUTES AND SCHOOLS

PETROBRAS UNICAMP

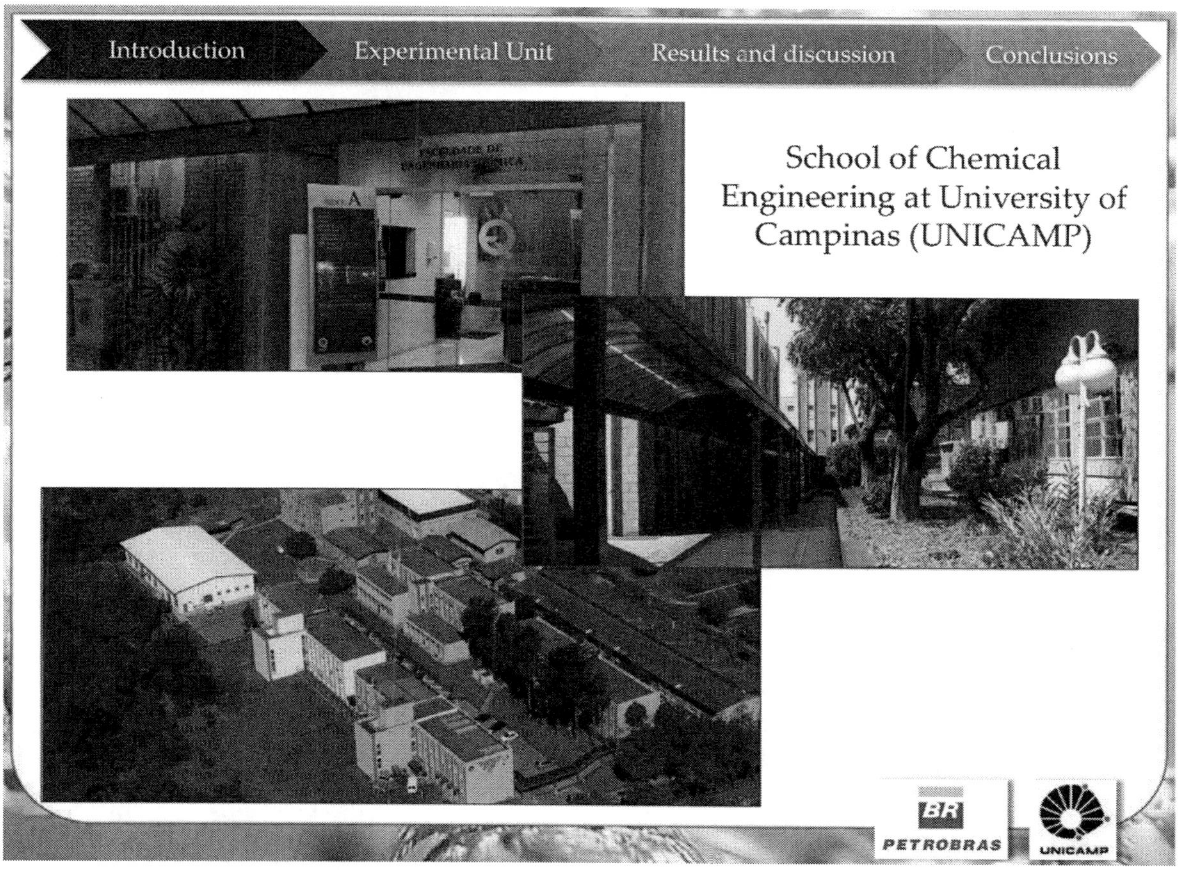
Introduction
Experimental Unit
Results and discussion
Conclusions
School of Chemical Engineering at University of Campinas (UNICAMP)
PETROBRAS
UNICAMP

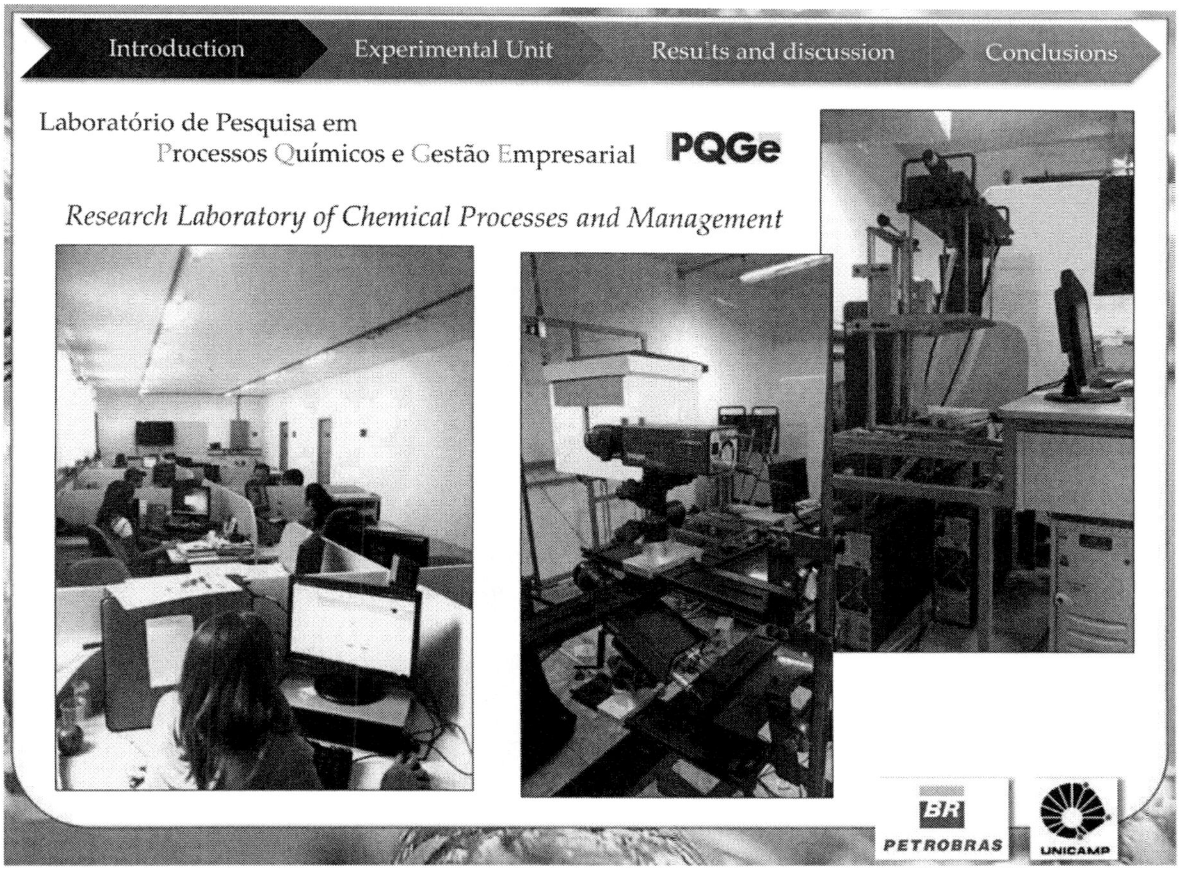
Introduction
Experimental Unit
Results and discussion
Conclusions
Laboratório de Pesquisa em Processos Químicos e Gestão Empresarial
PQGe
Research Laboratory of Chemical Processes and Management
PETROBRAS
UNICAMP

Introduction | Experimental Unit | Results and discussion | Conclusions

Laboratório de Pesquisa em
Processos Químicos e Gestão Empresarial **PQGe**

*Research Laboratory of Chemical Processes and Management*

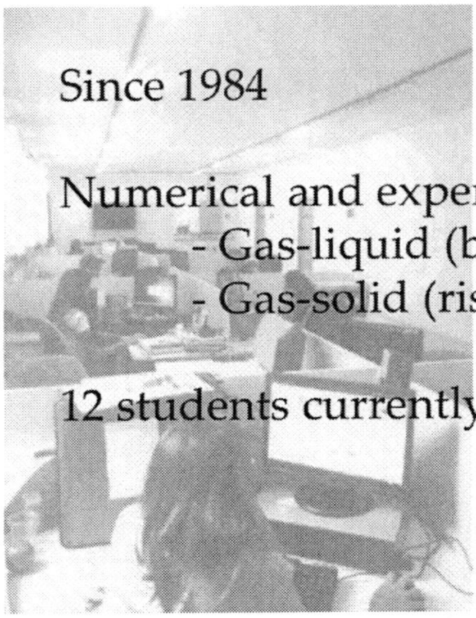

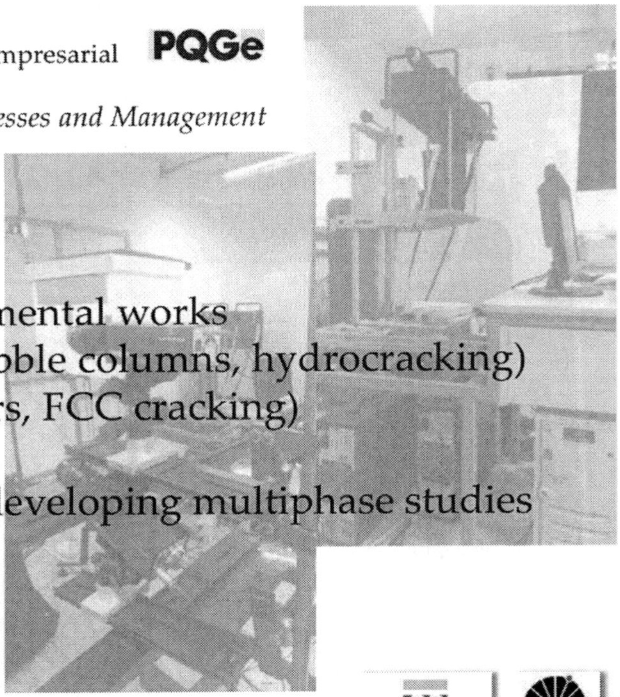

Since 1984

Numerical and experimental works
- Gas-liquid (bubble columns, hydrocracking)
- Gas-solid (risers, FCC cracking)

12 students currently developing multiphase studies

Introduction | Experimental Unit | Results and discussion | Conclusions

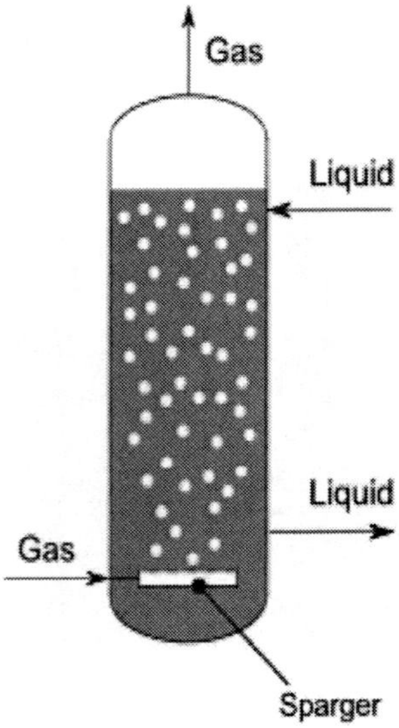


•VISCOSITY EFFECT :
BUBBLE FORMATION PROCESS AND THE TENDENCY TO COALESCENCE

•SOLID INFLUENCE : BUBBLE FORMATION, RISE BUBBLE VELOCITY, POROSITY, FLOW REGIMENS , LIQUID VELOCITY PROFILES AND MASS TRANSFER PHENOMENA.

•IMAGINE TECHNIQUES TO STUDY THE HYDRODYNAMICS:
- ➢Particle Image Velocimetry (PIV)
- ➢Stereo Particle Image Velocimetry (Stereo-PIV)

Proving a 2D-3C velocity field.

Introduction | Experimental Unit | Results and discussion | Conclusions

This work consists in the study of:

three-phase flow in a cylindrical bubble column

1.5 L/min gas flow rate

Fluorescent Stereo-PIV
CFD

Water
FCC particles
air
→ comparison →
CMC solution
FCC particles
air
→ Validation →
CFD Models

Liquid viscosity (CMC solution)

Load solid (FCC particles)

bubble column hydrodynamic characteristics

PETROBRAS UNICAMP

Introduction | Experimental Unit | Results and discussion | Conclusions

# Experimental Setup

Stereo-PIV
Bubble Column

PETROBRAS UNICAMP

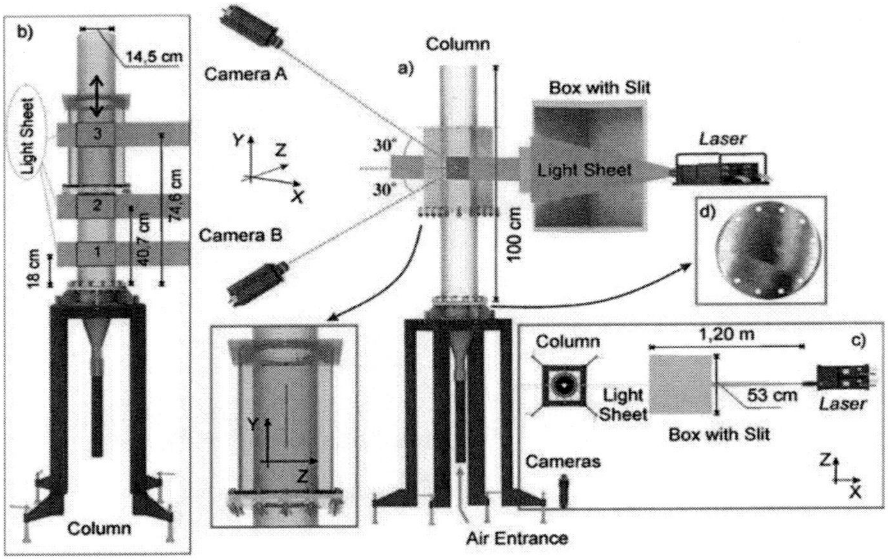


*Figure 1: Experimental unit facility*



**BUBBLE COLUMN**

| Experimental Conditions | | |
|---|---|---|
| Liquid Phase | Water (25°C) | |
| | CMC Solution 0,25% wt | |
| Gas Phase | Air Compressed | |
| Solid Phase | FCC Particles | |
| | 10 g | 15 g |
| Initial Liquid High | 80 cm | |
| Distance Measurement | 18 cm | |
| | 40.7 cm | |
| | 74.6 cm | |

**BUBBLE COLUMN**

$$\frac{\mu-\mu_0}{\mu_0-\mu_\infty}=\frac{1}{(1+(\lambda\dot{\gamma})^a)^{1-n/a}}$$

*Carreu -Yasuda model*

| Parameters | |
|---|---|
| $\mu_0$ | 0.2107 (Pas) |
| $\mu_\infty$ | 0.001 (Pas) |
| $\lambda$ | 0.9612 (s) |
| a | 0.3292 |
| n | 0.6534 |

Haake Rheostress (RS1) rheometer with a shear stress range ($\tau$) of 0.001 Pa-300 Pa

PETROBRAS UNICAMP



# PIV SYSTEM

2 CCD Camera: 1376X104 *Pixel*
*Imager Intense Cross- Correlation – Lavision Model*

- ❑ 4000 double images ~ 10 minutes of experiment

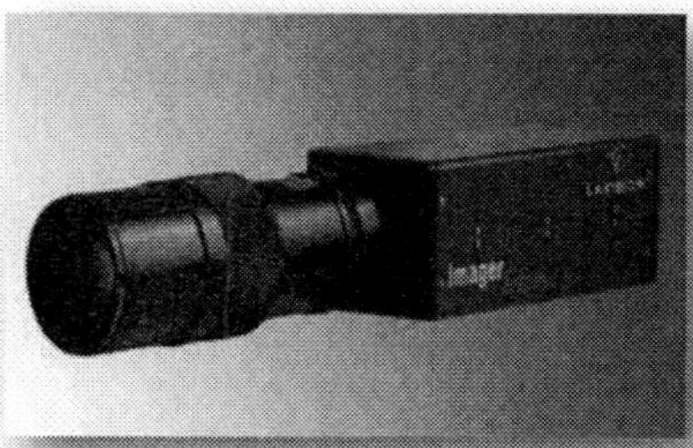

Table 1. Principal Parameters of Stereo-PIV setup

| Parameter | Value |
|---|---|
| Time inter-frame (dt) | 5000 μs |
| Maximum particle image displacement | 15-20 pixels |
| Acquisition frequency | 4.92 Hz |
| Pixel size | 6.45 μm |
| Laser power | 25-40% |
| Particle image diameter | 2-3 pixels |
| Image resolution | 10 pixels/mm |
| Field of view | 14.5 x 10 cm |
| Calibration error | ≈0.8 pixels |

*LASER*: Nd:YAG.
Wavelength of 532 nm.

Tracer Material:
carboxi-acrylate with dye-rhodamine ($C_{28}H_{31}N_2O_3Cl$).
Mean diameter : 15 μm
Specific mass: 1100 kg $m^{-3}$

PETROBRAS UNICAMP

Introduction | Experimental Unit | **Results and discussion** | Conclusions

# Results

<u>Water</u>

FCC catalyst

Air

PETROBRAS UNICAMP

Introduction | Experimental Unit | **Results and discussion** | Conclusions

## Vector field for liquid velocity Vy at 732mm

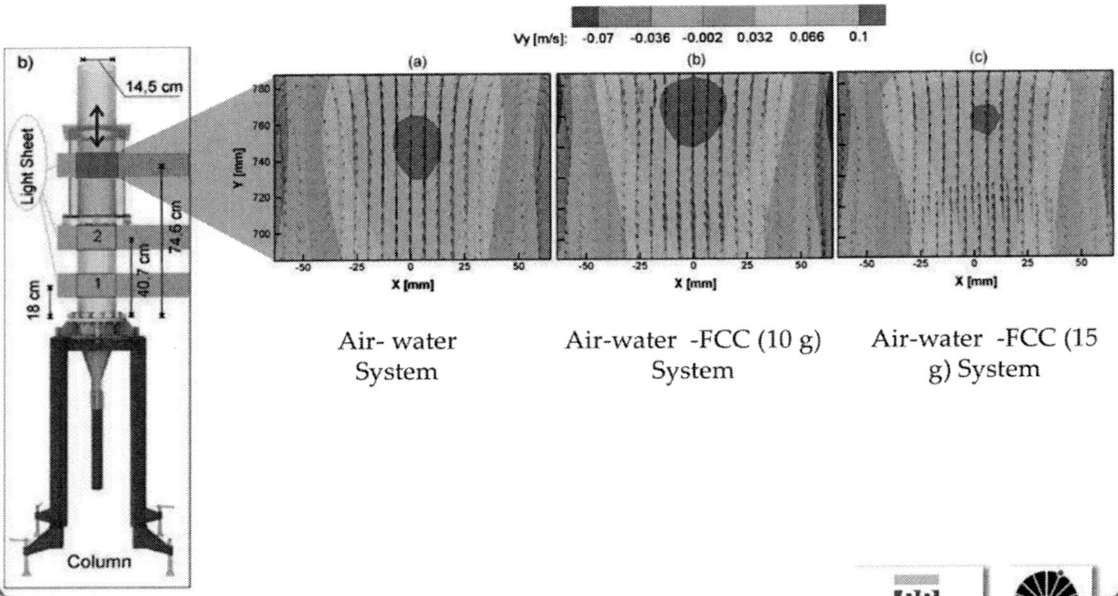


Air- water System | Air-water -FCC (10 g) System | Air-water -FCC (15 g) System

PETROBRAS UNICAMP

## Vector field for liquid velocity Vy at 732mm

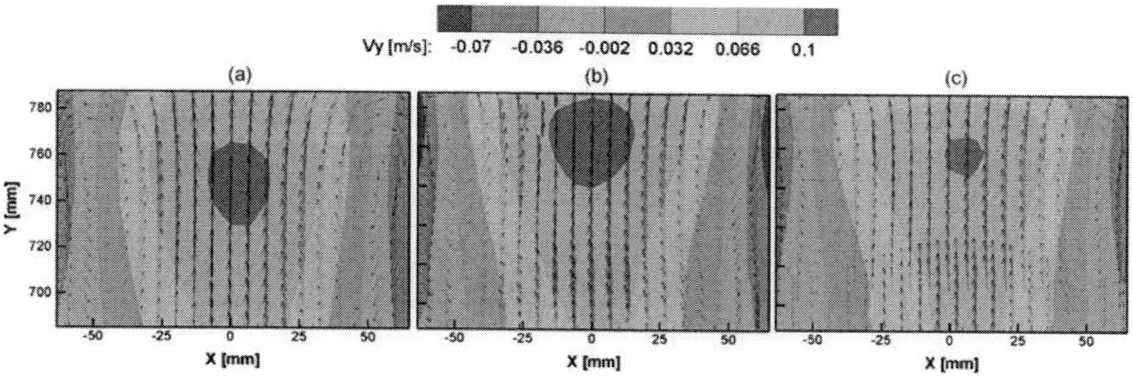


Air- water System

Air-water -FCC (10 g) System

Air-water -FCC (15 g) System





## Vector field for liquid velocity Vy at 392 mm

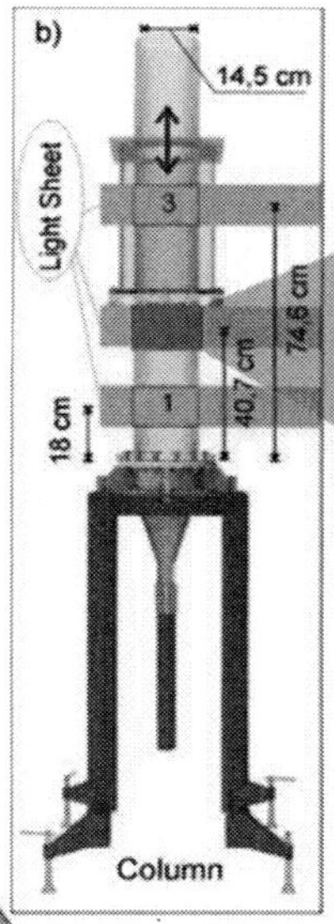


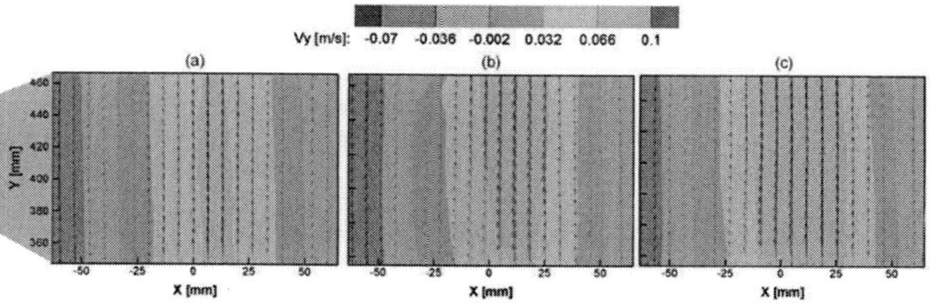


Air- water System

Air-water -FCC (10 g) System

Air-water -FCC (15 g) System

Introduction | Experimental Unit | Results and discussion | Conclusions

## Vector field for liquid velocity Vy at 392 mm

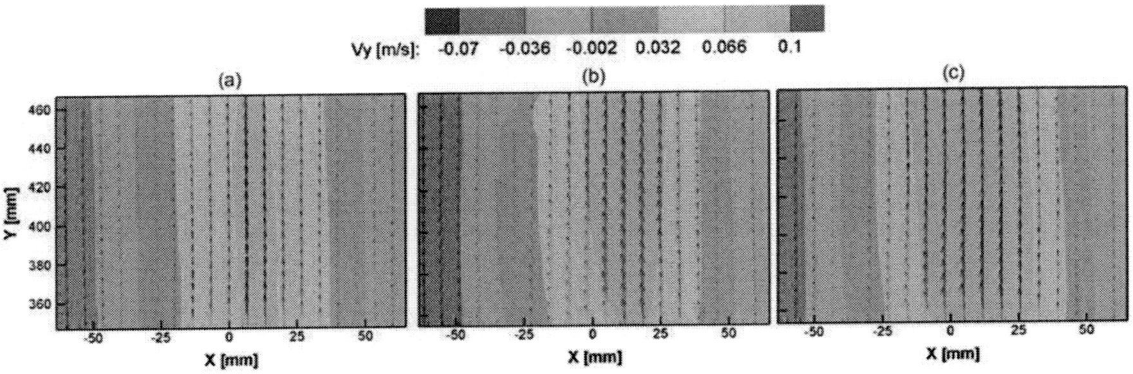


Air- water System

Air-water -FCC (10 g) System

Air-water -FCC (15 g) System

PETROBRAS UNICAMP

Introduction | Experimental Unit | Results and discussion | Conclusions

## Vector field for liquid velocity Vy at 165 mm

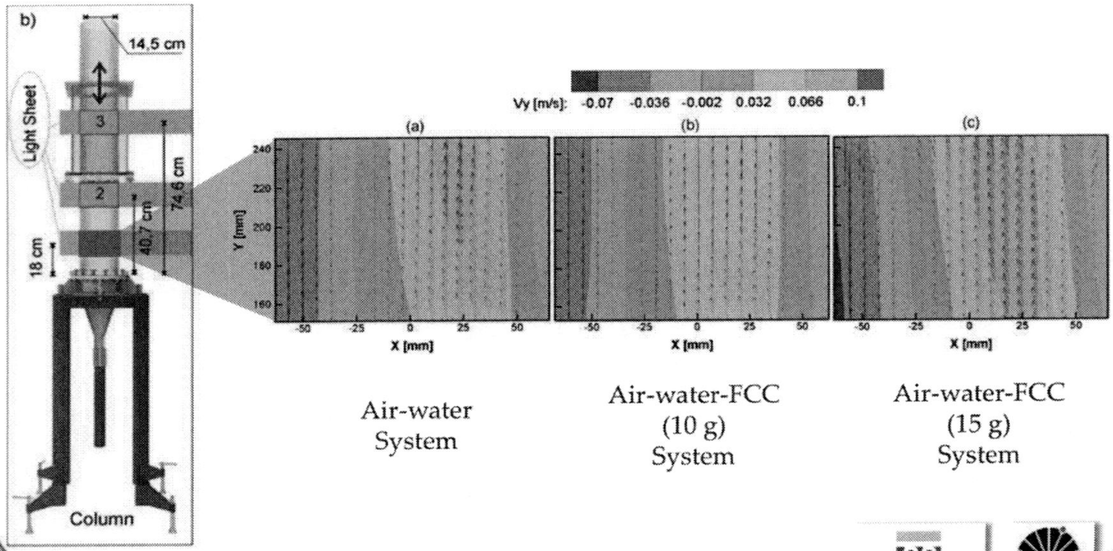


Air-water System

Air-water-FCC (10 g) System

Air-water-FCC (15 g) System

PETROBRAS UNICAMP

## Vector field for liquid velocity Vy at 165 mm

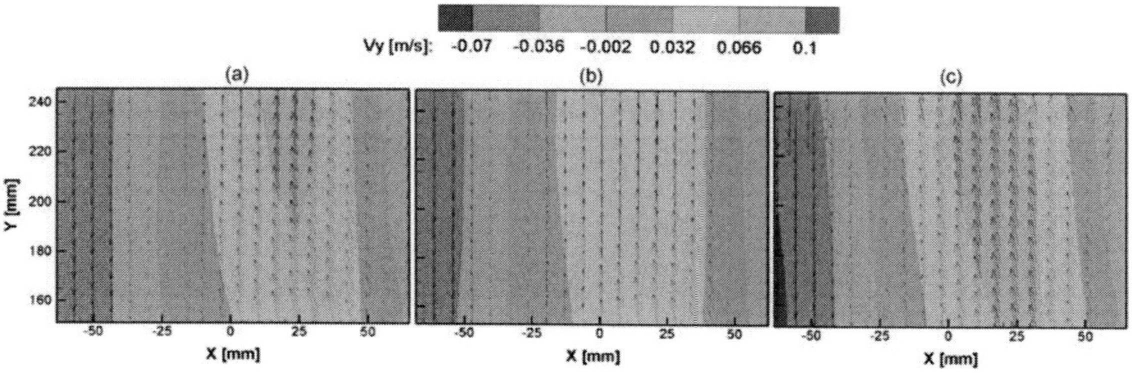


Air-water System

Air-water-FCC (10 g) System

Air-water-FCC (15 g) System

PETROBRAS UNICAMP



## Axial liquid mean velocity profile Vy

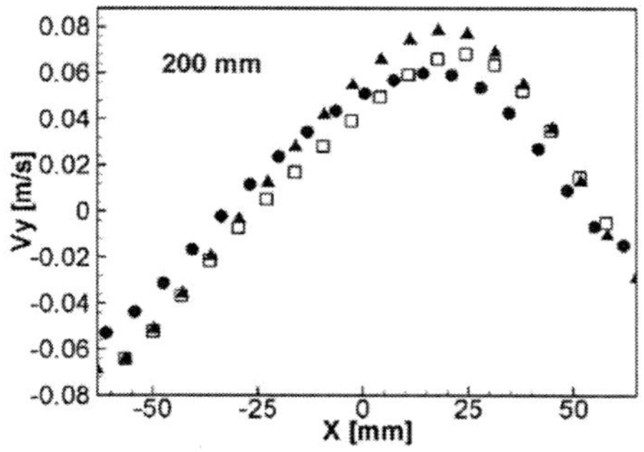


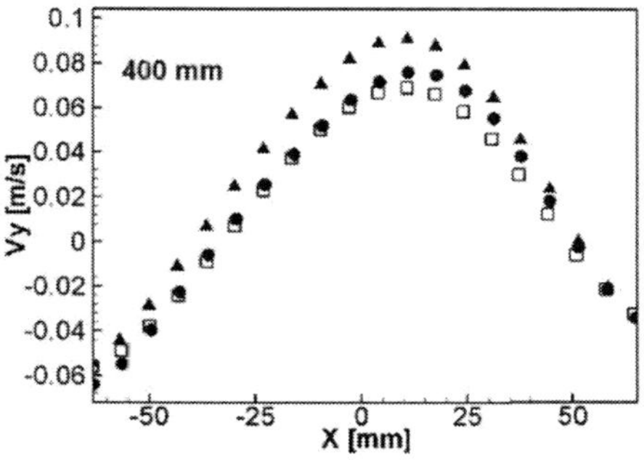


Water
Gas
FCC catalyst

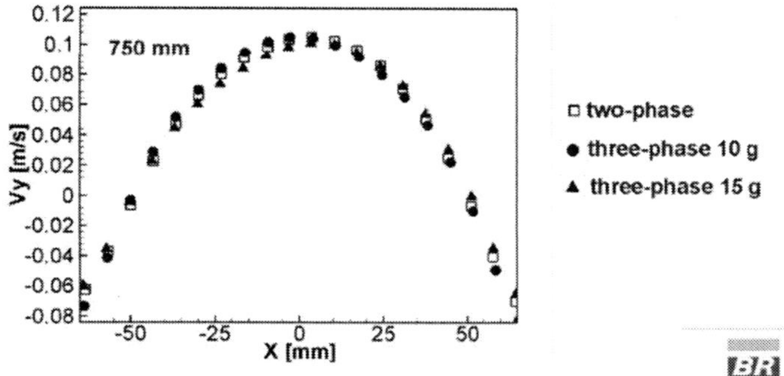


PETROBRAS UNICAMP

Introduction | Experimental Unit | Results and discussion | Conclusions

<u>CMC solution</u>

FCC catalyst

Air

PETROBRAS UNICAMP

Introduction | Experimental Unit | Results and discussion | Conclusions

## Vector field for liquid velocity Vy at 165 mm

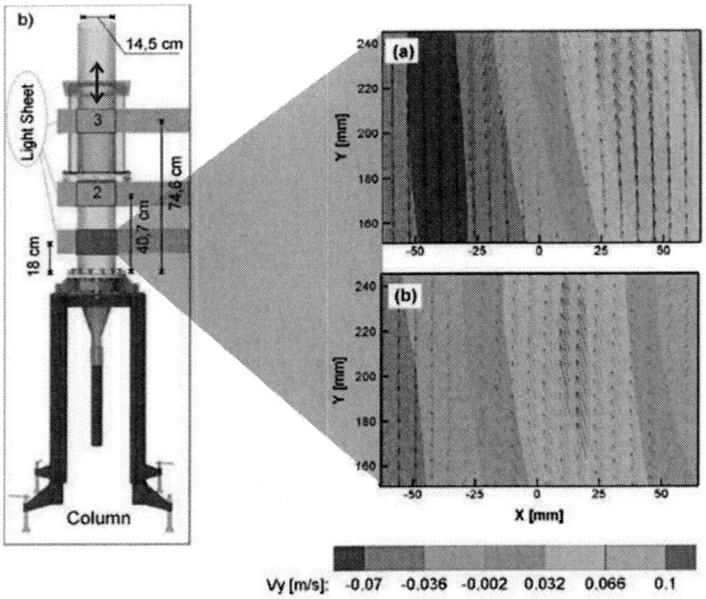


Air
CMC solution
(25% wt)

Air
CMC solution
(25% wt)
FCC particles
(15g)

PETROBRAS UNICAMP

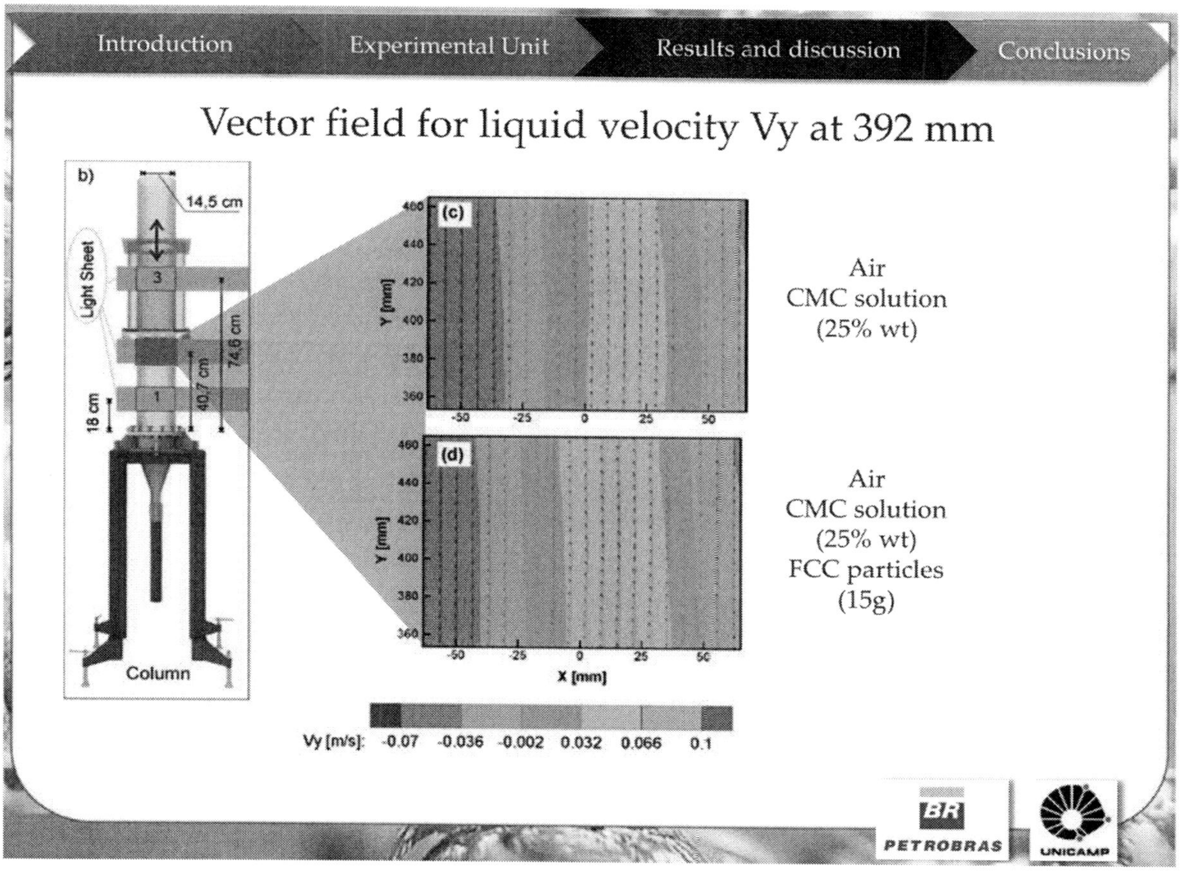
Introduction
Experimental Unit
Results and discussion
Conclusions
Vector field for liquid velocity Vy at 392 mm
b)
14,5 cm
Light Sheet
3
1
18 cm
40,7 cm
74,6 cm
Column
(c)
Y [mm]
(d)
X [mm]
Vy [m/s]: -0.07 -0.036 -0.002 0.032 0.066 0.1
Air
CMC solution
(25% wt)
Air
CMC solution
(25% wt)
FCC particles
(15g)
PETROBRAS
UNICAMP

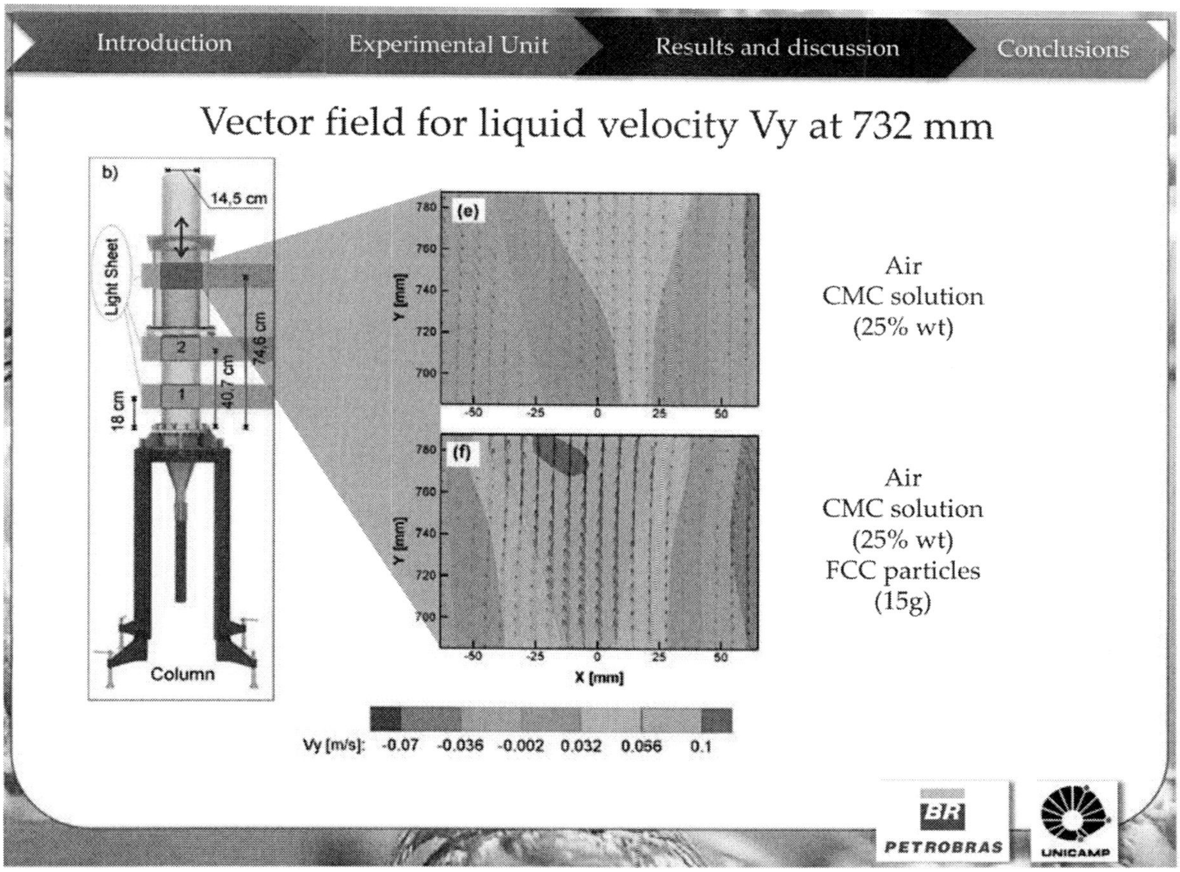
Introduction
Experimental Unit
Results and discussion
Conclusions
Vector field for liquid velocity Vy at 732 mm
b)
14,5 cm
Light Sheet
2
1
18 cm
40,7 cm
74,6 cm
Column
(e)
Y [mm]
(f)
X [mm]
Vy [m/s]: -0.07 -0.036 -0.002 0.032 0.066 0.1
Air
CMC solution
(25% wt)
Air
CMC solution
(25% wt)
FCC particles
(15g)
PETROBRAS
UNICAMP

Two-phase

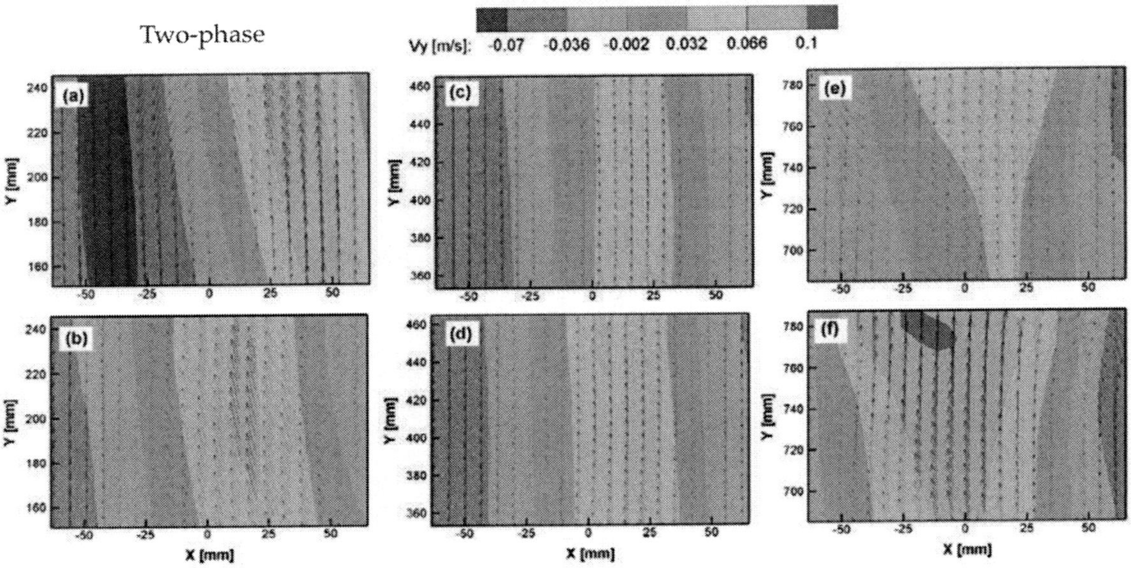


Three-phase

PETROBRAS

UNICAMP

Introduction | Experimental Unit | Results and discussion | Conclusions

# Axial liquid mean velocity profile Vy

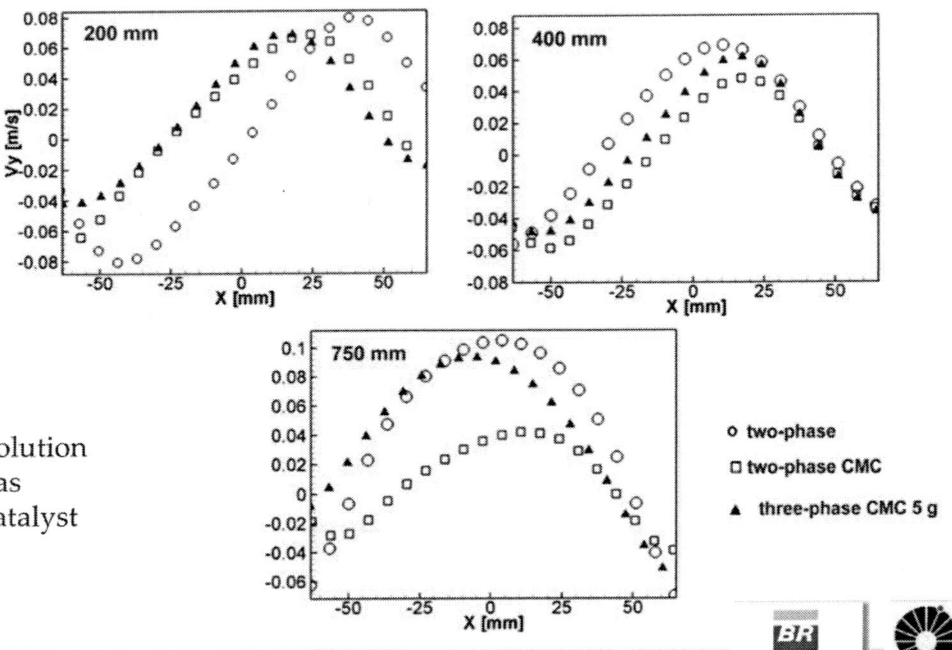


CMC solution
Gas
FCC catalyst

PETROBRAS

UNICAMP

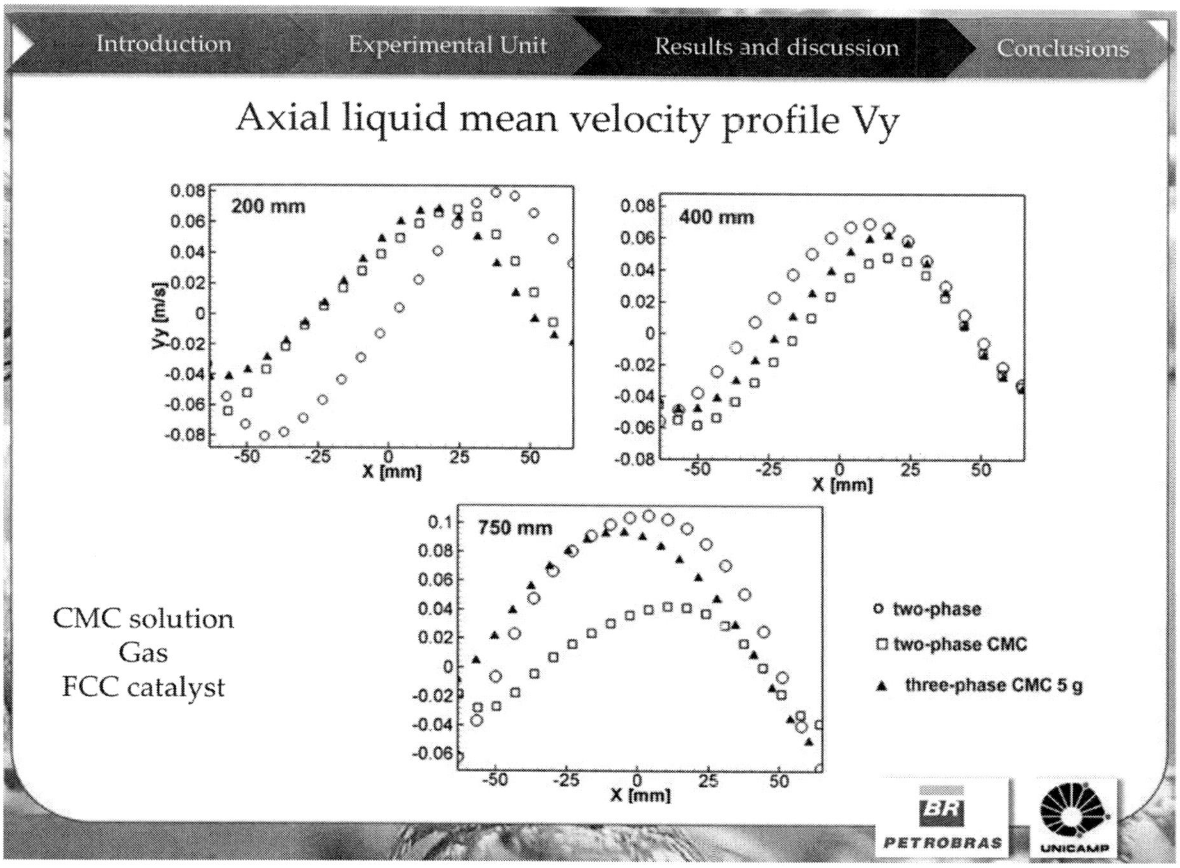
Introduction
Experimental Unit
Results and discussion
Conclusions
Axial liquid mean velocity profile Vy
200 mm
400 mm
750 mm
Vy [m/s]
X [mm]
CMC solution
Gas
FCC catalyst
two-phase
two-phase CMC
three-phase CMC 5 g
PETROBRAS
UNICAMP

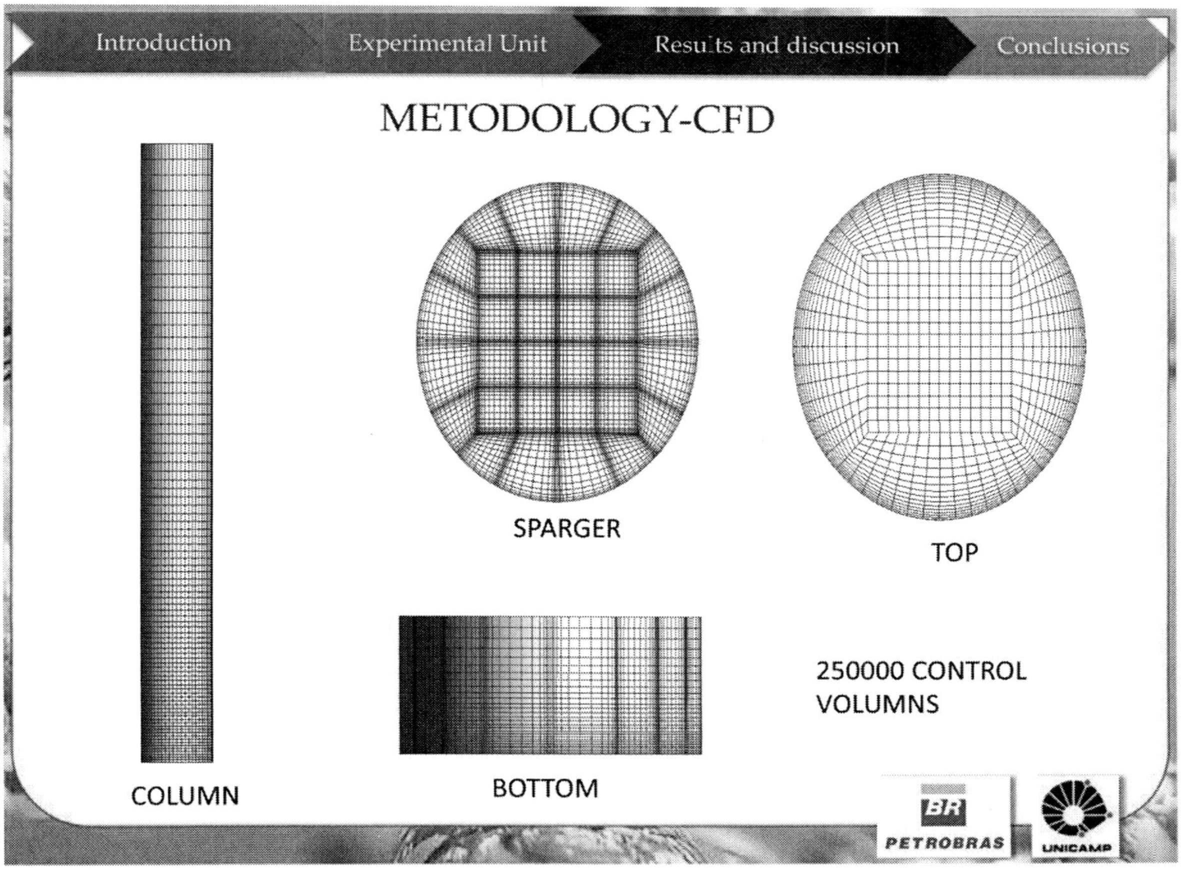
Introduction
Experimental Unit
Results and discussion
Conclusions
METODOLOGY-CFD
SPARGER
TOP
250000 CONTROL VOLUMNS
COLUMN
BOTTOM
PETROBRAS
UNICAMP

Introduction | Experimental Unit | Results and discussion | Conclusions

# Mathematical approach (CFD)

**Eulerian Multiphase Approach**

**Silva Jr *et al*,(2013)**

- RANS model
  Reynolds Averaged Navier-Stokes
- Turbulence model RNG k- ε + bubble induced turbulence (SATO e SEKOGUCHI (1975))
- Drag force Gas-Liquid
  Izhii e Zuber (1979)
- Drag force Liquid-Solid
  Wen e Yu (1966) modified by Jia et al. (2007)
- Liquid phase viscosity
  Carreu-Yasuda Model

PETROBRAS UNICAMP

Introduction | Experimental Unit | Results and discussion | Conclusions

| Fluid | Density (Kg/m$^3$) | Viscosity (Kg/ms) | Superficial Tension (mN/m) |
|---|---|---|---|
| Air | 1.255 | $1.789*10^{-5}$ | |
| Water | 997 | 0.001 | 72 |
| CMC 0.25%wt | 1002 | Carreu-Yasuda | 71.22 |
| FCC Particles | 1800 | $1.8*10^{-5}$ | |

| ***Initial conditions*** | |
|---|---|
| Water | $H_l$=80cm, $\alpha_l$=1, $\alpha_g$=0, $\alpha_s$=0.00042 |
| CMC | $H_l$=80cm, $\alpha_l$=1, $\alpha_g$=0, $\alpha_s$=0.00042 |
| T=25C, $P_{rel}$=0 | |
| ***Boundary Conditions*** | |
| Inlet | $u_g$= 1,19 m/s, $\alpha_l$=1, $\alpha_g$=$\alpha_s$=0 |
| Oulet | $\alpha_g$=1, $\alpha_l$=$\alpha_s$=0 Type: opening |
| Wall | Gas and liquid: non slip<br>Solid: free slip |

PETROBRAS UNICAMP

Introduction | Experimental Unit | Results and discussion | Conclusions

# Experimental and numerical Axial liquid mean velocity profile Vy

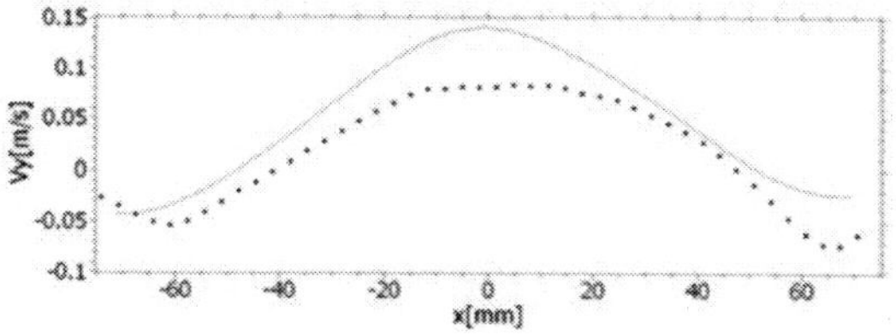


10 l/min
H1

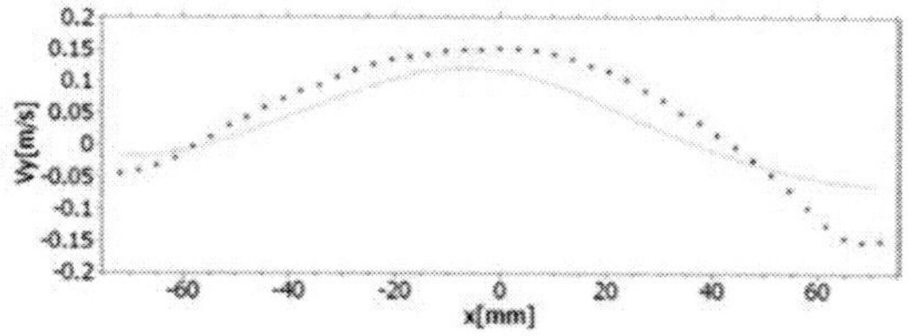


10 l/min
H2

0,25 % wt CMC
solution
Gas

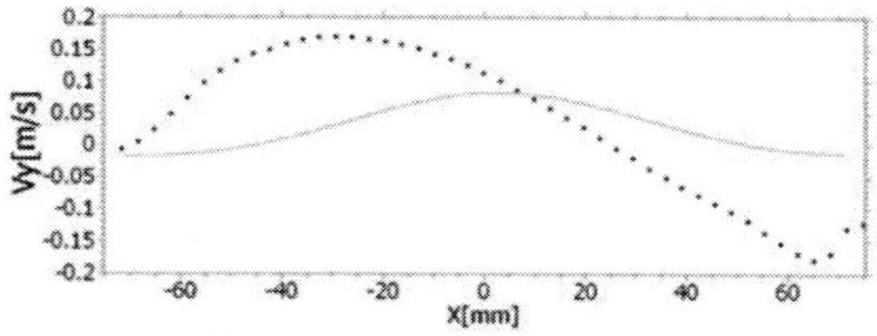


10 l/min
H3

PETROBRAS UNICAMP

Introduction | Experimental Unit | Results and discussion | Conclusions

# Experimental and numerical Axial liquid mean velocity profile Vy

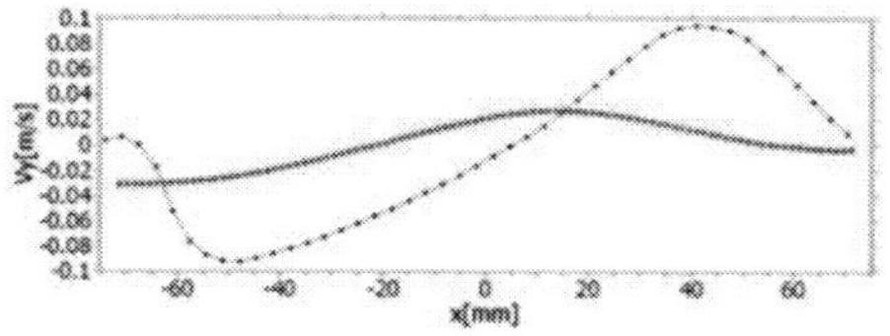


1,5 l/min
H1

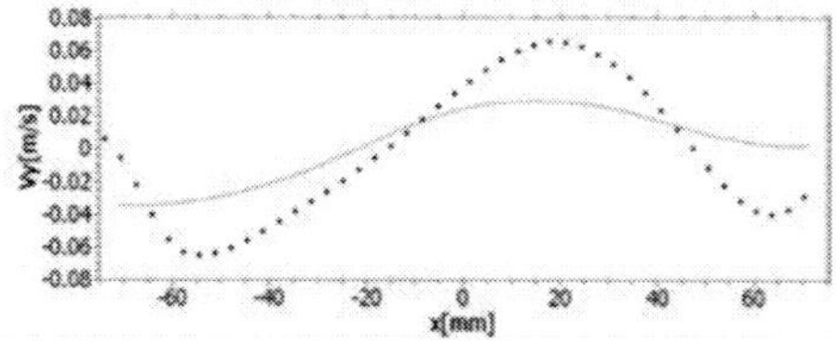


1,5 l/min
H2

0,25 % wt CMC
solution
Gas

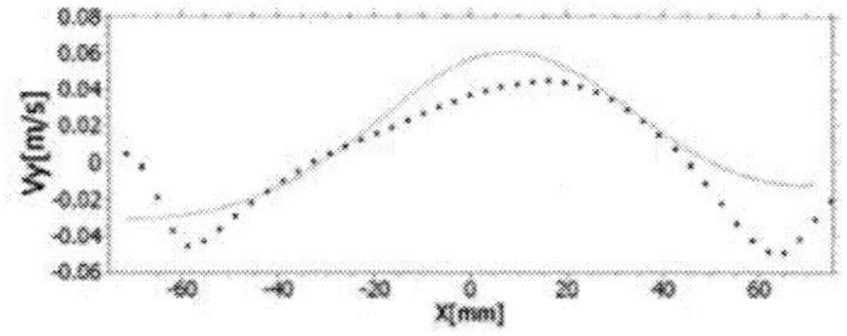


1,5 l/min
H3

PETROBRAS UNICAMP

Introduction | Experimental Unit | Results and discussion | Conclusions

The experiments performed with Fluorescent Stereo- PIV provided data for fluid dynamics characterization. The experimental results show typical axial velocity profiles of the liquid, upward flow in the core region and a down-flow near the wall.

It was observed that the solid effect is significant in the column bottom where the flow is not developed; while at the top of the column it is not for the small concentration of the solid added into the system. Moreover, the liquid viscosity effect (two-phase flow) is higher. It can be observed that the liquid velocity decrease as the liquid viscosity increase, because the bubbles have an extra resistance that allow the movement upward. Besides, the viscosity promotes the bubble coalescence that affects the liquid velocity. However, the studied effects are not significant by the fact that the superficial gas velocity and solid load be low.

Good agreement between numerical and experimental data.

PETROBRAS UNICAMP

# THANK YOU!


***Acknowledgements***

The authors acknowledge the support of:

PETROBRAS UNICAMP PQGe

# 458071 CFD Simulation for Bubble Column with and without Internals of Fischer Tropsch (FT) Synthesis

**Abstract:**

Very limited studies either experimentally or numerically have investigated the influence of vertical internals on the hydrodynamics properties and flow pattern in bubble column equipped with the internals. Therefore, in this study, the impact of vertical internal on time -averaged cross-sectional gas hold distribution and their radial profiles have experimentally and numerically addressed in 6-inch bubble column without and with internals of the air-water system at churn turbulent flow regime (15 and 20 cm/s). In this work, thirty dense Plexiglas internals have arranged uniformly with the hexagonal configuration that covers 25% of the total cross-sectional area of the column, which similar to those employed in Fischer-Tropsch (FT) synthesis. The time-averaged gas holdup distribution and their radial profile at different superficial gas velocities have simulated in a full 3D, transient, Euler/Euler framework by utilizing the computational fluid dynamic CFD. The 3D simulation results of the time-averaged cross-sectional gas holdup distributions and their radial profiles have been validated with benchmarking data measured by using gamma-ray computed tomography (CT) technique for the same system. The result and the findings will be presented in the presentation.

# 461485 Evaluating Mixing Performance in a Stirred Tank Reactor with Different Impeller Geometries and Reactor Size By CFD

**Abstract:**

Stirred tank reactors are ubiquitous in the pharmaceutical industry and their mixing efficiency is paramount on the rate of heat and mass transport processes and on the reaction rate as well. Traditional experimental measurement techniques are restricted to the limited number discrete of points were the measurements are taken. Moreover, measurements are difficult in reactors functioning at extreme conditions involving cryogenic temperatures or high pressures.

The numerical approach circumvents this experimental limitation and allows for the flow pattern to be evaluated to its full extent. Furthermore, CFD minimizes the risk associated with scaling up processes involving complex geometries where classical empirical correlations might not ensure the mixture components are well mixed. CFD also allows to assess the influence of several parameters and operational conditions to define an operational design space, very much aligned with the Quality-by-Design (QbD) approach. This enables an enhancement of process knowledge and to easily troubleshoot potential issues occurring in the production trains.

The objective of this work is to evaluate the mixing performance of a stirred tank reactor by assessing the distribution of a solute starting from a stagnant layer at the bottom of the reactor. Different impeller geometries and reactor scales were tested using Computational Fluid Dynamics (CFD). These simulations were implemented to study a single phase incompressible fluid by means of a RANS-type finite-volume (FV) method. Turbulence was modeled by the two-equation standard high-Re k-epsilon turbulence model, as implemented in OpenFOAM, in tandem with both a Moving Reference Frame (MRF) and Arbitrary Mesh Interface (AMI) approaches to account for the effects due to impeller rotation. New solvers were developed to determine the concentration field of the tracer solute: a steady state solver for MRF and a transient state one for AMI. The passive scalar transport approach was applied to simulate the transport of a scalar quantity in an incompressible fluid flow. The primary assumption of this approach is that the solutes are non-reactive, with no influence on the flow field, acting therefore as numerical tracers in the flow and thus are used to evaluate the mixing efficiency of the impeller.

The steady state and transient flow fields were calculated for given flow conditions using new solvers based on simpleFoam (Semi-Implicit Method for Pressure-Linked Equations) and pimpleDyMFoam (merged PISO-SIMPLE) , respectively. These new solvers include the transport equation for a passive scalar that is solved along with the velocity and pressure fields. The new solvers were used to compare the suitability of either approach (MRF and AMI) in the description of mixing of the passive scalar in typical pharmaceutical reactors. In this study a 95 % mixing uniformity was considered as an indication of a well mixed system.

# 466987 CFD Simulation on Assessment of Biologic Product Mixing Performance in Single Use Technology


**Abstract:**

The biopharmaceutical industry and has undertaken radical change in the design and operation of many bioprocess. These changes provide more robust process and increased production flexibility. The benefits of Single Use Technology (SUT) for equipment train flexibility and reduced cleaning logistics are now widely acknowledged. This trend has resulted in rapid growth of SUT(equipment) manufactures and product offerings.

In order to select the right SUT from multiple vendors for protein drug product mixing operation, a CFD simulation was applied to assess the impact of the size and shape of tank and impeller,, fill level,Â and agitation rate on the flow and mixing behavior. An approach was developed by introducing two concepts: the highest accumulated shear rate and mixing time safety factor. These parameters, respectively, quantify the acceptable boundaries of shear and mixing time validated within the operational ranges established for a commercial product based on product specific stability profile.

The model simulations were used to assess the SUTs mixing performance, the mixing time, highest accumulated shear rate, and range of acceptable agitation rate for three different SUTâ€™s vendors. Compared with the existing mixing operation for the commercial product, the mixing performance was classified and the decision was made to select the optimalÂ SUT for the product being considered.Â

# 467302 Characterization of Lab and Plant Reactors Using Fast Competitive Reactions and a Predictive Model

**Abstract:**

Scale-up of reaction and crystallization conditions often involves assumptions about equivalence of mixing between lab and plant. Scale-down of processes for lab studies require similar assumptions. A question that often arises in pharma applications is how small can the lab equipment be, so that the results translate to scale. With modern lab synthesis workstations providing good control of addition rates and temperature in reactors that resemble their larger scale equivalents, scale-up and scale-down work sometimes takes place in lab systems with volumes as low as 100 mL. It is possible using mixing-sensitive test reactions to quantify the liquid mixing environment in vessels at any scale and to thereby identify lab and plant conditions that are more quantitatively equivalent.

The fast competitive reaction system involving neutralization and ester hydrolysis (a so-called 'Bourne Reaction', ref 1) among other techniques has been used by Pfizer to study mixing characteristics in several common lab reactors used for routine development of chemical processes and also for a series of plant reactors. The final product distribution after running the reaction in fed batch mode reflects the history of relative mixing and reaction rates in the plume or reaction zone that forms near the feed location.

In general, a critical feed time can be identified and longer feed times represent micromixing-controlled conditions. It is possible using a predictive model containing both kinetics and mixing parameters to estimate the product distribution as a function of feed time and mixing energy dissipation rate. This latter parameter, often referred to as epsilon, with units of W/kg, is of interest for a wide range of mixing phenomena. Using the predictive model together with measured product distribution, it is possible to estimate the corresponding epsilon.

We will present and discuss results for different reactors and operating conditions and also how best to use predictive models to design and analyse the experiments, which require a good dynamic range of selectivity to be informative. We will compare mixing performance at operating conditions that are commonly used in the lab and plant. Finally, we will extract some guidelines towards obtaining equivalent mixing between lab and plant reactors.

1. Investigation of micromixing in stirred tank reactors using parallel reactions, John R. Bourne , Shengyao Yu, Ind. Eng. Chem. Res., 1994, 33 (1), pp 41–55

# 469803 Computational and Experimental Determination of the Hydrodynamics in a Scaled-Down Pharmaceutical Vessel Provided with Angle-Mounted Impellers

**Abstract:**

In most industrial applications, stirred tanks and reactors are typically provided with baffles to improve their mixing characteristics. However, in a number of pharmaceutical production facilities unbaffled vessels are commonly used. The absence of baffles is preferred in such cases because it reduces the potential for contamination and makes cleaning the vessel between batches easier. However, the lack of baffles also has a negative impact on the system's hydrodynamics since it often results in poor mixing of the batch, especially if the impeller is centrally placed, since the liquid in such a system is subject to a strong tangential flow, but low axial and radial flows. This hydrodynamic regime is highly undesirable if effective top-to-bottom liquid recirculation is required, and especially if a second phase (e.g., solid particles, immiscible liquids) must be dispersed and incorporated into the liquid bulk. For this reason, impellers in unbaffled pharmaceutical vessels are typically placed off-center, and, additionally, they may be mounted on a shaft that is angled with respect to the vessel vertical centerline. This lack of symmetry introduces some degree of "baffling" effects, in that it promotes stronger vertical recirculation of the liquid, and improves the suspension of settling solids and the incorporation of floating solids.

Despite their industrial relevance, especially for the pharmaceutical industry, extremely limited information is available on the hydrodynamics of vessels provided with angle-mounted impellers. Therefore, in this work, experimental and computational tools were used to determine the hydrodynamics of fluids in a Plexiglas, custom-made, scaled-down version (diameter: 304 mm) of an industrial vessel with an elliptical bottom and provided with two angle-mounted (by 5 off the vertical) A-310 Lightnin impellers. The system was operated under different operating conditions in order to replicate the mixing characteristics of the industrial system. Computational Fluid Dynamics (CFD) was used to quantify the hydrodynamics of this system under different geometric configurations (such as liquid level), agitation speeds, and for different fluid rheologies. In all cases a multiple reference frame (MRF) computational approach was used and turbulence was modeled using the k-ε method. In addition, Particle Image Velocimetry (PIV) was used to experimentally determine the velocity flow field in water for some of these configurations in order to validate the CFD predictions, thus providing guidance on the optimal operation of these industrial systems.

The results obtained here indicate that there is substantial agreement between the CFD predictions and the PIV experimental results. The flow in the vessel appears to be very complex. The axial pumping action of the impellers produces a downward flow impinging the bottom of the vessel resulting in flow splitting and in the formation of a less well-mixed zone near the vessel bottom that persists even when the impeller velocity is substantially increased. This zone could be the preferential location for the sedimentation of settling solids in the liquid.

# 471722 Experimental Study of Wet Milling in a Rotor-Stator Mixer

**Abstract:**

While wet milling of active pharmaceutical ingredients (API) and other crystalline substances has commanded considerable attention, there are few fundamental studies on which to base mill design, scale up and operating strategies. Often an in-line rotor-stator mixer, fed from and recirculated to a well-mixed holding tank, is used to achieve the ultimate particle size or to precondition a feed to a higher intensity device such as a ball mill. In this study, a systematic framework is developed and employed with definitive experiments performed in a Silverson L4R rotor-stator mixer, to investigate the effect of milling conditions, mixer geometry, and particle physical properties on ultimate particle size and particle breakage kinetics. By varying operating conditions, mixer geometry and crystal mechanical properties, it is possible to discriminate among various breakage mechanisms (fluid shear, particle-particle collisions, particle-wall collisions, etc.) to develop mechanistic relations to aid in process scale up and to serve as inputs to population balance and other models for the evolution of the crystal size distribution.

Based on the literature, it was reasoned that along with size and shape, stress intensity factor (or fracture toughness) and hardness are the important physical properties affecting breakage rate. To vary these properties, several crystalline materials were milled in antisolvents (e.g. sugar in isopropyl alcohol) to allow discrimination between mechanical size reduction and solubility effects. Collision rates were varied by changing particle concentration; and coatings and small geometric changes were used to consider interactions with solid surfaces. Rotor speed and volumetric throughput were adjusted to independently vary energy input and mill head residence time. Crystal hardness and stress intensity factor were measured with a Hysitron TriboIndenter. Particle size distributions were measured by several methods. A Lasentec focused beam reflectance measurement (FBRM) probe and Particle View Microscope (PVM) were placed in the holding tank to continuously monitor particle size and suspension uniformity. Grab samples, acquired in the holding tank at well-defined time intervals, were analysed using a Horiba laser diffraction instrument and by microscopy with an automated image analysis technique. The results from the various measurement techniques will be

compared and discussed. Mechanistic models that account for particle properties and milling conditions will be presented for the various fracture mechanisms. The mechanistically correlated experimental results will be discussed in detail, as well as application to process scale up. One intended use of the correlations is as mechanistic inputs (breakage kernels) to population balance models.

# 474214 Scale-up Rules for the Generation of Solid-Stabilized Emulsions

**Abstract:**

Effects of scale-up on the generation of solid stabilized emulsions (SSEs) were investigated experimentally. Considering unbaffled systems of 0.25, 5, and 35 Liters with geometrical similarities, emulsions were prepared using off-centred pitched blade and Rushton turbines. Standard scale-up criteria allowing for the calculation of impeller speed at a larger scale based on $N_{large} = N_{small}(D_{large}/D_{small})^a$ and commonly used for surfactant-stabilized emulsions (energy dissipation, Weber number, Reynolds number, impeller tip speed and circulation time) were applied for concentrated and diluted formulations of SSEs. Silicone oil (50 cSt) was used as the dispersed phase in water and solid glass microspheres ($d_{32} \sim 3\mu m$) as stabilizers. Emulsions were characterized by means of droplet size measurements using Mastersizer 3000 from Malvern. Sauter mean diameter ($d_{32}$), $dv_{10}$ and $dv_{90}$ were used to compare the resulting emulsions. A well-established decolorization technique was used to evaluate mixing times. The results from the experiments showed that none of the standard scale-up criteria is suitable for scaling-up Pickering emulsions processes. Weber number, Reynolds number and tip speed as scale-up criteria lead to larger diameters with increasing scale while the opposite tendency is observed with circulation time and energy dissipation. Results tend to indicate that constant drop size distribution can be obtained using an intermediate between the energy dissipation rate and the tip speed suggesting that the suitable scale-up exponent a of $N_{large} = N_{small}(D_{large}/D_{small})^a$ would be between 2/3 (energy dissipation rate) and 1 (impeller tip speed). The presentation will present the results and their interpretation.

# AUTHOR INDEX

# AUTHOR INDEX

**AIChE**
120 Wall Street, FL 23
New York, NY 10005-4020

ISBN 978-1-5108-3442-2

CURRAN ASSOCIATES INC.
proceedings.com